Advances in Anatomy Embryology and Cell Biology

Vol. 124

Udo M. Spornitz

The Functional Morphology of the Human Endometrium and Decidua

With 80 Figures

Springer-Verlag Berlin
Heidelberg GmbH

PD Dr. Udo M. Spornitz
Anatomisches Institut,
Pestalozzistrasse 20, CH – 4056 Basel

ISBN 978-3-540-54519-4 ISBN 978-3-642-58099-4 (eBook)
DOI 10.1007/978-3-642-58099-4

Library of Congress Cataloging-in-Publication Data. Spornitz, Udo M. The functional morphology of the human endometrium and decidua / Udo M. Spornitz. p. cm. — (Advances in anatomy, embryology, and cell biology; vol. 124) Includes bibliographical references and index. ISBN 978-3-540-54519-4
1. Endometrium — Anatomy. 2. Decidua — Anatomy. I. Title. II. Series: Advances in anatomy, embryology, and cell biology; v. 124. [DNLM: 1. Decidua — anatomy & histology. 2. Decidua — physiology. 3. Endometrium — anatomy & histology. 4. Endometrium — physiology. W1 AD433K v. 124 / WP 400 S764f] QL801.E67 vol. 124 [QM421] 574.4 s — dc20 [612.6'2] DNLM/DLC for Library of Congress.

Originally published by Springer-Verlag Berlin Heidelberg New York 1992

Typeset by Best-set Typesetter Ltd., Hong Kong

Contents

1 Introduction

1.1 Historical Perspective

In the nineteenth century, knowledge of the events leading to ovulation, fertilization, and implantation was very limited, so much so that Seiler (1832), in his book *The Uterus and the Human Egg*, wrote: "... in the left ovary the first signs of fertilization, namely a Graaf vesicle could be seen. The right ovary shows proof of a second successful copulation: a fresh scar from the ovulated egg and the beginning of a corpus luteum." In fact all nineteenth century authors strictly divide the female cycle into two phases: the menstrual period and the intermenstruum (cf. Hitschmann and Adler 1908). The generally accepted histology of the endometrium in those days was that of the late proliferative phase. Deviations from this were considered to be pathological (Von Ebner 1902). As Gebhard (1899) expressly put it: "As a rule, it can be said that in the mature woman the endometrial glands run straight; an irregular course of the glands is to be regarded as pathological." The same author describes the changes occurring during the secretory phase of the cycle as "endometritis glandularis" which he believed to arise from a local nutritional disturbance. The uterine stroma was believed to be lymphoid (Toldt 1877), and the uterine glands were compared to the crypts of Lieberkühn (Von Ebner 1902). The first authors to show that the endometrium passed through four different phases during the course of a cycle were Hitschmann and Adler (1908), whose results were later confirmed by other authors, particularly Schröder in 1915. The basis for a cycle-related dating of the endometrium was, however, not given before the late 1930s when Rock and Bartlett (1937) published a significant paper on the endometrium. Their results were later refined by Noyes et al. who firmly established the techniques of "dating the endometrial biopsy" (1950).

The development of techniques applicable to biology and medicine made electron microscopy in the early 1950s a powerful tool which allowed deeper insight into the cell. Among the first authors to study the endometrium with the electron microscope were Lanzavecchia and Morano (1958); Wetzstein (1958); Morano et al. (1959); and Vecchietti and Morano (1959). Even now, the findings reported in these early papers are amazing, particularly when one considers the slow penetration rate of fixatives used in electron microscopy and the speed with which autolytic changes occur.

1.2 Hormone Action on the Endometrium

Our view of the uterus as a possible target organ for ovarian hormones originated at the beginning of this century in the case of progesterone (Fraenkel 1903; Meyer 1911), and in the late 1930s in the case of estrogen (Astwood 1938). It is now clear that the cyclic variations in endometrial morphology are due to the action of progesterone and estrogen which is mediated through cytoplasmic receptor molecules (Clark et al. 1977). The number of receptors varies during the course of a cycle and is dependent upon the amount of circulating hormone and the length of exposure to the hormone (Flickinger et al. 1977). The formation of the progesterone receptor protein involves the action of estrogen (Leavitt et al. 1977) in the form of a "priming action", prior to which progesterone does not exert its full effects. On the other hand, there is an inverse relationship between the amount of circulating progesterone and the number of progesterone receptors in that increasing amounts of progesterone reduce the number of progesterone receptors (Vu Hai et al. 1977). Progesterone apparently increases the inactivation rate of the receptor. Moreover progesterone also suppresses the formation of the estrogen receptors, the numbers of which are drastically reduced during the secretory phase (Flickinger et al. 1977).

The formation of hormone receptors appears to be strongly influenced by local uterine mechanisms; ectopic endometriotic implants do not generally respond to circulating hormones as well as the eutopic endometrium does (Schweppe et al. 1984; Schweppe and Kindermann 1988). Even within the eutopic endometrium local factors seem to play a major role in the mediation of hormone action illustrated by the fact that it is in particular the stratum functionale which reacts to circulating hormones and is subsequently shed during menstruation. The basalis remains to a large extent unaltered by ovarian hormones. Morphological changes of the endometrium during the proliferative phase are brought about mainly by estrogen, while changes during the secretory phase of the cycle are induced by progesterone. Since ovarian hormones produce only very small changes on the basalis, this paper focuses primarily on the stratum functionale.

1.3 Ultrastructural Features of the Endometrium

1.3.1 Proliferative Phase

1.3.1.1 General Aspects

The light microscopic morphology of the endometrium, in particular the changes which occur during the proliferative and secretory phase of the cycle, is well known and is, therefore, not described in this introduction. The most significant results of electron microscopic investigations are, however, reported since they provide a basis for the results reported in the factual part of this book.

The proliferative phase of the female cycle is generally subdivided into an early, middle, and late proliferative phase. The effects produced by ovarian hormones on the stroma differ greatly from the effects on the epithelial cells; therefore, they will be dealt with separately.

1.3.1.2 Glandular Epithelium

Early Proliferative Phase (Days 5–7). The glandular epithelial cells possess a relatively straight cell membrane in the lateral as well as in the basal parts of the cells (Verma 1983). The apical membrane is augmented by microvilli, the baseline of which also runs straight (Wynn and Harris 1967). The nucleus is located in a central, basal area and the nuclear membrane shows no major undulations. The cells have a distinct nucleolus (Cornillie et al. 1985). The Golgi apparatus usually consists of only a few supranuclear dictyosomes with a narrow lumen (Cavazos et al. 1967). Lysosomal elements are distributed over the entire cytoplasm as small, mostly autophagic vacuoles. Endoplasmic reticulum (ER) and polysomes are few in number (Verma 1983). There are only a few mitochondria present which show no topographical relationship to the ER as they do in later stages (Verma 1983). The surface epithelium has a morphology very similar to that of glandular epithelial cells. Ciliated cells are occasionally found intermingled with regular surface epithelial cells.

Middle Proliferative Phase (Days 8–10). The lateral cell membrane shows pronounced undulations (Verma 1983). The number of microvilli at the apical cell membrane remains basically unaltered but the baseline of the microvilli follows a convex course (Cornillie et al. 1985). The nuclei have left their basal position and moved a little further towards the center of the cells. The nuclear membrane begins to show undulations. Mitotic figures which can be seen during the entire proliferative phase, have drastically increased in number (Verma 1983). ^{3}H-Thymidine studies have shown that the rate of tissue proliferation is at its highest between days 8 and 10 (Ferenczy et al. 1979b). Compared with the early proliferative phase the amount of ER has increased and can frequently be found in close contact with the mitochondria which are also more numerous. In the subnuclear region free ribosomes and polysomes are quite abundant (Wynn and Harris 1967). The number of lysosomes has decreased (Cornillie et al. 1985). Both the cells of the glandular epithelium and the cells of the surface epithelium have grown taller. The number of ciliated cells within the surface epithelium has increased (Ludwig and Metzger 1976a). The ciliated cells can also be found in gland openings, almost one-third of their length extending into the glands (More and Masterton 1975).

Late Proliferative Phase (Days 11–14). The glandular epithelial cells are now columnar and pseudostratified (Ferenczy and Richart 1974). The cytoplasm and the mitochondrial matrix have become more electron dense (Gordon 1975). The amount of free ribosomes, polysomes, and rough endoplasmic reticulum (RER) has decreased, particularly in the basal, subnuclear part of the cells (Cornillie et al. 1985), where occasional glycogen particles are present (Wynn and Woolley 1967). The Golgi apparatus is even more enlarged and can be found almost exclusively in the apical, supranuclear part of the cells (Verma 1983). The mitochondria are almost entirely surrounded by RER (Verma 1983). A slight enlargement of the mitochondria is evident around days 13–14 (Gordon 1975). In the upper parts of the glands and in the surface epithelium the number of ciliated cells has reached a maximum (Ludwig and Metzger 1976b).

1.3.1.3 Stroma

During the proliferative phase of the cycle, changes in stromal cells are not very pronounced; therefore, it seems unnecessary to distinguish between the early, middle, and late proliferative phase, as has been done with the glandular cells. The typical stromal cell is spindle shaped and possesses only a very narrow rim of cytoplasm (Wynn and Woolley 1967; Ferenczy and Richart 1974; Spornitz et al. 1984). This gives rise to the histological phenomenon "naked nuclei" (Wynn and Woolley 1967). The stromal cell is described as a pluripotent mesenchymal cell by Dallenbach-Hellweg (1975). Wynn and Woolley (1967), however, describe it as a normal connective tissue cell. Both authors, nevertheless, agree on the fact that the stromal cell hardly changes its morphology during the proliferative phase of the cycle. Wienke et al. (1968) describe changes around ovulation which lead to the formation of what they call a fibroblast-like cell. Particularly in the upper parts of the stratum functionale, these authors find that the smooth endoplasmic reticulum (SER) and RER are grossly augmented. Moreover, the number of vesicles present in the Golgi areas is increased (Sengel and Stoebner 1970). During the whole of the proliferative phase stromal cells are engaged in mitotic activities (Wynn and Woolley 1967; Ferenczy et al. 1979b). Towards the end of the proliferative phase, intracellular membrane-bounded collagen fibrils are also found in stromal cells (Wienke et al. 1968).

1.3.2 Secretory Phase

1.3.2.1 Glandular Epithelium

1.3.2.1.1 General Aspects

The criteria for the histological dating of endometrial tissue are well known and need not be repeated here (Noyes et al. 1950). Because of the minute size of specimen prepared for electron microscopy, endometrial dating is in most cases difficult, if not impossible. Electron microscopy can, however, refine the dating of histologically predated tissue. The early to middle secretory phase is characterized on the ultrastructural level by the appearance of three morphological entities: (a) the nucleolar channel system (NCS), (b) giant mitochondria, and (c) subnuclear glycogen. All of these structures appear to be dependent upon the postovulatory changes in hormonal levels, particularly the surge of progesterone. For this reason they are called "the postovulatory triad" (Cornillie et al. 1985).

1.3.2.1.2 Nucleolar Channel System

Among the magnitude of subcellular structures which can only be detected with the aid of electron microscopy is the NCS. This organelle was described as early as 1960 by Dubrauszky and Pohlmann, who called it a nucleolus with a peculiar grating. Several years later, Clyman (1963) described the same organelle as a "nucleolar basket". Ancla et al. (1964) simply call it "corpuscules intranucléaires," stating that "ils evoquent des images de nids d'abeilles". Terzakis (1965) was the first to call this organelle the "nucleolar channel system". This is the term most often used, although the relation between the NCS and the nucleolus has not been fully elucidated. The NCS, in fact, sometimes appears to be located outside the nucleolus (More et al. 1974).

The NCS appears to be induced by progesterone, specifically by the β-acyl group present in the D ring of this steroid (Kohorn et al. 1972). In tissue culture only this specific functional group of progestogens was able to induce the formation of the NCS. Gestagens with an acetyl group in the 17-β position or an acyl group in the 19-β position were not able to induce the formation of the NCS (Kohorn et al. 1972). The application of estrogen during the 3rd to 8th postovulatory days results in a regression of the NCS (Gordon et al. 1973).

As a progesterone-dependent structure, the NCS is present in the endometrium for only a few days (More and McSeveney 1980). Clyman (1963) said it was present from day 16 through 26 of the cycle, during which time it is formed and disintegrates. It therefore follows that the fully developed NCS is present for an even shorter period of time. The maximum number of fully developed NCS are, in fact, present in the glandular epithelial cells around day 19. Therefore, it is understandable why investigations carried out at 5-day intervals (days 5, 10, 15, 20, 25) (Cavazos et al. 1967; Cavazos and Lucas 1973) failed to detect the NCS. Despite early knowledge of its existence (Dubrauszky and Pohlmann 1960), the NCS has not been satisfactorily described. It has even been given a different morphological description by the same authors at different times (e.g., More et al. 1974, 1975; More and McSeveney 1980); their perception of the NCS ranges from a simple helical arrangement to a supercoiled helix. According to the 1974 paper, for example, the NCS consists of two tubules which run in nine turns around the center of the coil, ending blindly in the nucleoplasm. Most authors agree that the NCS consists of tubules which have a diameter ranging between 60 and 100 nm. The tubules are embedded in a matrix of granular material, surrounding an electron lucent center (cf. Wynn 1978). The granular material appears to consist of ribonucleoprotein (RNP) and ribosomes; the ribosomes have the diameter typical for ribosomes (15 nm) (Gordon 1975). The fully developed NCS measures approximately 1000 nm (Moricard and Moricard 1964; More and McSeveney 1980). It is, in most cases, in close proximity to the nuclear membrane, especially to an invagination of the nuclear membrane (Dubrauszky and Pohlmann 1960; Clyman 1963; Terzakis 1965; Wynn and Woolley 1967; Spornitz 1987).

Occasionally up to three NCS can be found inside the same nucleus (Terzakis 1965; Sengel and Stoebner 1970; Armstrong et al. 1973).

1.3.2.1.3 Giant Mitochondria

Giant mitochondria were first described in the human endometrium by Dubrauszky and Pohlmann (1960), Nilsson (1962b), and Gompel (1964), whose findings have since been confirmed in the majority of studies performed during the postovulatory phase of the cycle. Together with subnuclear glycogen and the NCS, the giant mitochondria form the postovulatory triad (Cornillie et al. 1985). In contrast to the regular mitochondria (pre- and postovulatory) which measure approximately 0.3 × 1 µm, giant mitochondria can be as large as a nucleus, i.e., 7 µm (Merker et al. 1968). Most of them, however, do not measure more than 3–4 µm. The increased size, compared with regular mitochondria, is not due to mitochondrial swelling, but is the outcome of genuine growth, evidenced by the fact that the number of cristae also increases (Sengel and Stoebner 1970a). Thus, there are two types of mitochondria present during the postovulatory phase, namely giant mitochondria and regular mitochondria. Most authors agree that the giant mitochondria are a feature of the second half of the cycle (Dubrauszky and Pohlmann 1961; Gompel 1964; Gordon 1975; Cornillie et al. 1985), which seem to develop as a result of rising titers of progesterone. Wynn and Woolley (1967) observed giant mitochondria before ovulation had occurred, but they do not, however, state how the relevant day of the cycle had been determined. It might well be that ovulation had either already occurred, or that they had investigated one of the rare cases of luteinized unruptured follicle syndrome (LUF) (Dmowski et al. 1980). With LUF the ultrastructure of the endometrium is identical to that of a normal cycle (Cornillie et al. 1985). Although the relationship between progesterone and giant mitochondria does not appear to be as specific as that between progesterone and the NCS, progestogenic hormones are clearly responsible for the induction of giant mitochondria (Kohorn et al. 1972).

1.3.2.1.4 Glycogen

Light microscopic studies have shown that large amounts of glycogen are not apparent in the glandular epithelial cells before ovulation has occurred. The first signs of glycogen detected by light microscopy are the basal vacuoles which result from the dissolution of glycogen during preparation. These vacuoles appear on day 15 or 16 of the cycle (Noyes et al. 1950). Electron microscopic investigations have shown that glycogen deposition already starts 2 days earlier (Wynn and Harris 1967). It has also been reported that glycogen is even deposited during the preovulatory phase (Armstrong et al. 1973; Gordon 1975; Verma 1983). The deposition of glycogen in the glandular epithelial cells does not appear to be related to either the intake of carbohydrates or to physical exercise (Gregoire et al. 1973). Initially, glycogen is deposited in the subnuclear region. The nuclei are consequently dislocated into the apex of the cells, which gives the cells a stratified appearance. As the cycle progresses the glycogen moves from the basal to the apical region of the cells and the nuclei are relocated to their original basal position (Wilborn and Flowers 1979). In the rat uterus glycogen synthesis can be induced through administration of estrogen alone (Williams and Provine 1966). In the human endometrium the induction of glycogen synthesis involves the priming action of estrogen which must, however, be followed by progesterone (Hughes et al. 1969). Furthermore, progesterone also induces glycogenolysis around the middle of the secretory phase; initially, this is counterbalanced by the increased glycogen synthesis brought about by the second estrogen peak (Hackl 1971). Estrogen seems to exert its action through activation of glycogen synthetase, while progesterone co-activates glycogen synthetase and then appears to activate phosphorylase. Thus, eventually glycogenesis as well as glycogenolysis occur side by side in the same cell (Hughes et al. 1963). Glucose-6-phosphate-dehydrogenase activity (G-6-PDH) is greatly diminished through progesterone (Jelinek et al. 1978).

1.3.2.2 Stroma

1.3.2.2.1 Predecidual Cells

To date, electron microscopic studies have paid little attention to the postovulatory stroma. A reason for this might be that the changes which occur during the early secretory phase of the cycle (Wynn and Woolley 1967) are too minute to serve as a basis for exact dating. The majority of stromal cells are fibrocyte-like cells which, as already mentioned, are believed to be either simple connective tissue cells (Wynn and Woolley 1967) or pluripotent mesenchyme-like cells (Dallenbach-Hellweg 1975). During the early secretory phase these cells multiply mitotically. During the middle and late secretory phase they differentiate into predecidual cells (Wienke et al. 1968). This involves development of a synthetic apparatus, including mitochondria, RER, and Golgi apparatus (Wienke et al. 1968; Wynn and Woolley 1967; Sengel and Stoebner 1970a,b). Around day 20 of the cycle glycogen appears in the cells in large amounts and is subsequently utilized by the cells (Wienke et al. 1968), and by day 23 most of the stromal cells in the upper part of the stratum functionale have been transformed into predecidual cells (Wynn and Woolley 1967). These cells are characterized by their size (larger than those in the early secretory phase), well-developed organelles, and the presence of intracellular collagen vacuoles (Sengel and Stoebner 1970). Without subsequent implantation of a blastocyst the predecidual cells are desquamated during menstruation, together with other cells of the stratum functionale (Bartelmez 1957; Dallenbach-Hellweg 1975).

1.3.2.2.2 Endometrial Granulocytes (K Cells)

During the secretory phase of the cycle a second cell type is present in the endometrium in greater numbers. The morphology of this cell differs greatly from the other stromal cells. It is the same cell type that is present in the basal decidua during the first trimester of pregnancy and it occurs in numbers equal to that of genuinely decidual cells (Hellweg 1959). This second type of cell contains light microscopically visible granules (German, *Körnchen*) which led to the term *Körnchenzelle*, abbreviated in some English papers to K cells (Hamperl and Hellweg 1958). Other authors call these cells "endometrial granulocytes" (Bulmer and Sunderland 1983). Both terms are, however, destined to cause misunderstanding; K cell is also used in immunology where it stands for killer cell and the term granulocyte is used in hematology to denote a type of leukocyte. This misleading terminology might be one of the reasons why, inspite of their outstanding morphology, very little attention has so far been paid to these cells. Another reason might be that these cells, first described as early as 1904 (Marchand), have been given different characteristics by different authors e.g., they are both metachromatic and nonmetachromatic (Ulesco-Stroganoff 1908; Asplund and Holmgren 1947; Cordier and Snoeck 1949). Some authors have claimed that K cells are identical to the metrial gland cells of rodents (Larkin and Cardell 1971; Peel and Bulmer 1977); whereas Hellweg (1954) describes them as being totally different.

1.4 The Action of IUDs on the Endometrium

1.4.1 General Considerations

From experience gained with inert IUDs, i.e., IUDs made from polyethylene or polypropylene, the idea of bioactive IUDs was born. The inert IUD exerts its contraceptive effect by means of a foreign body reaction (FBR) (Moyer and Mishell 1971), the magnitude of which depends upon the shape and particularly the size of the IUD (Tatum 1972). The larger the contact area of the IUD with the endometrium, the more pronounced the FBR. Unfortunately a pronounced FBR not only increases the contraceptive efficacy of thc IUD but also secondary effects such as cramps, bleeding, or even expulsion of the IUD (Piotrow et al. 1979). In order to reduce the side effects the size of IUDs was reduced and this resulted in a higher Pearl index (pregnancies per 100 woman/years) (see Wagner 1976). Bioactive substances were then added to the IUD in an attempt to increase the contraceptive efficacy without increasing the side effects. This led to the development of copper IUDs (Zipper et al. 1969; Van Os et al. 1980) and progesterone IUDs (Pharriss et al. 1974). The copper IUD which is manufactured in a large variety of forms (T, 7, bow, etc.) comprises a copper wire wound mostly around a vertical arm. There is a reciprocal relationship between the number of failures (pregnancies) and the size of the surface of the copper wire on the IUD: the larger the copper surface, the lower the Pearl index (Van Os et al. 1980; Zipper 1984).

The progesterone IUD is T shaped and made from ethylene-vinyl-acetate. In the vertical arm of the IUD there is a reservoir which contains 38 mg progesterone (Martinez-Manautou 1975). The progesterone is delivered at a daily rate of approximately 65 µg (Pharriss et al. 1974). Because of the size of the IUD in relation to the size of the uterine cavity, delivery of progesterone from the IUD leads automatically to a gradient of progesterone concentration within the uterine cavity. This again leads automatically to a gradient of morphological response to the progesterone by the endometrium (Spornitz et al. 1983; Mall-Haefeli and Spornitz 1984).

The exact localization of biopsy material is of critical importance in the investigation of physiological, i.e., untreated endometrium. There are considerable morphological differences between biopsies taken from the area of the tubal openings and tissue from the corpus and fundus of the uterus (Spornitz et al. 1980; Van Lüdinghausen and Anastasiadis 1984). Even within tissue from the same area there are frequently islands of cells which do not respond properly to the circulating hormones (Shepard and Senturia 1977). This might result from an inhomogeneous distribution or inconsistent concentration of receptors for progesterone and estrogen (Tsibris et al. 1978); what ever the reason, it is clearly difficult to get consistent results on the basis of biopsies alone.

By the very nature of its structure and mode of progesterone delivery, the progesterone IUD can be expected to render inconsistent results. This is also the case with the copper IUD, due to the concentration gradient of released copper ions within the uterine cavity (Wagner 1976) – the endometrium presents a different morphological response to released copper ions, according to the concentration gradient.

1.4.2 Mode of Action

The mode of action of IUDs on the endometrium is primarily a surface phenomenon, resulting from direct contact of tissue with the IUD (Ludwig 1980; El-Badrawi and Hafez 1980; De Brux et al. 1981). According to Ludwig the mechanical effect of IUDs is based on three factors:

1. Actual localization within the uterine cavity (with respect to anteversion and anteflexion).
2. The constant movements of the IUD caused by uterine contractions.
3. The FBR.

Because of anteversion and anteflexion of the uterus, the IUDs generally exert more mechanical wear on the posterior wall than to the anterior wall of the uterus, impressions on the posterior wall are usually clearly visible to the naked eye (El-Badrawi and Hafez 1980; De Brux et al. 1981). Due to mechanical wear, there is frequently damage to the surface epithelium of the endometrium, which is accompanied by leukocytic infiltration. The remaining intact cells of the surface epithelium acquire a structural polymorphism, particularly with copper IUDs (Ludwig 1980; Ludwig and Metzger 1984), and occasionally micropolyp-like formations have been reported (Ludwig 1980; De Brux et al. 1981; Ludwig and Metzger 1984; Spornitz 1985). The diffusion of bioactive substances (copper ions, progesterone) leads to morphological changes even in tissue areas which are not in direct contact with the IUDs (Spornitz et al. 1980; Spornitz et al. 1983). Decidualization of the stroma is one of the most outstanding changes resulting from the use of progesterone IUDs (Ludwig and Spornitz 1977; Spornitz et al. 1980; Spornitz et al. 1984). The progesterone IUD also influences the glandular epithelium, as reflected by reduced secretion from and even atrophy of the cells (Spornitz et al. 1980; Mall-Haefeli and Spornitz 1984).

1.5 Decidua

1.5.1 Historical Perspective

After implantation, the predecidual cells of the late secretory phase develop into true decidual cells. The decidua can be divided into:

1. Decidua basalis (underlying the implantation site).
2. Decidua capsularis (between implantation site and lumen of the uterus).
3. Decidua parietalis (lining the entire uterine cavity, except for areas 1 and 2 above).

The decidua was first described as early as 1794 by William Hunter (cited in Ramsey 1973), although the term "decidua" itself was coined by Vesalius in the sixteenth century. At the beginning of the twentieth century it was discovered that the endometrium could be induced to form decidual tissue using various nonspecific stimuli (Loeb 1908). Loeb had already established that, in addition to the nonspecific stimulus, a functioning corpus luteum had to be present in order for the transformation of endometrial stroma into decidual tissue to take place. To distinguish artificially induced decidua from physiological decidua Loeb called the former the "deciduoma." The first deciduomata were induced in guinea pig uteri. Since then it has been shown that deciduomata can be induced in practically all of the investigated mammals. The applicable stimuli range from the purely mechanical traumatizing of the endometrium (Blandau 1961; De Feo 1963), through intraluminal injection of physiological substances such as histamine (Shelesnyak 1957), sodium bicarbonate (De Feo 1963), and prostaglandin (Sananes et al. 1981; Miller and O'Morchoe 1982), to the presence of unphysiological materials like oil droplets (Finn and Keen 1962) or IUDs (Ludwig and Spornitz 1977). With respect to the successful embryo transfer of in vitro fertilized ova, the perifocal decidualization resulting from mechanical traumatization during transfer represents a serious problem (Dallenbach-Hellweg et al. 1984).

1.5.2 Ultrastructural Features of the Decidua

The first ultrastructural investigation of human decidua was made by Lawn et al. (1971), who gave a detailed description of important elements of decidual tissue e.g., the presence of an extracellular coat of basal lamina (BL)-like material and the presence of strongly osmiophilic granules at the periphery of the cells; the presence of the former was later confirmed by Wynn (1974). Karyosomes were observed by Liebig and Stegner (1974), who interpreted the presence of these structures as evidence of the participation of decidual cells in steroid metabolism. Changes in the morphology of mitochondria were not reported by the above-mentioned authors. In contrast to these authors, however, Schmidt found considerable amounts of lipid in decidual cells (Schmidt et al. 1982). Using immunochemical methods the BL-like coat was found to contain laminin, fibronectin, and type IV collagen (Wewer et al. 1985).

1.6 Ectopic Endometrium and Ectopic Decidua

1.6.1 Ectopic Decidualized Tissue

Ectopic decidual cells were first described towards the end of the nineteenth century. Kinoshita found ectopic decidua in the ovary (1898) and Schmorl found ectopic decidual cells in the vicinity of the urinary bladder and in the area of the rectouterine cavity (1897). In the fallopian tubes decidual cells have been found during tubal pregnancy (Wist 1954; Rewell 1972). Rewell, however, cited an earlier paper by Mall (1915) who said: "we have examined with care all the walls (of the fallopian tubes from 117 cases of tubal pregnancy) and as yet have never found any tissue that could be considered as decidua." Ovarian decidual cells towards the end of pregnancy were described in detail by Herr et al. (1978, 1987). On the basis of these papers it can be said that ovarian decidual cells are very much like the genuine endometrial decidual cells in most respects. Since there are no further ultrastructural studies on ectopic decidual cells nothing more can be said about the nature of decidual cells in other locations.

1.6.2 Ectopic Endometrium

There appears to be a topographic relationship between the localization of ectopic decidua and endometriotic implants (Weller 1935; Herr et al. 1978). The presence of ectopic endometrium, generally called endometriosis, has not only been reported in numerous organs of the pelvis, but also in the lungs, the cecum, and even muscles of the arm (Dougherty and Anderson 1964; Kuzela and Speers 1985; Schmidt 1985; Kindermann 1988). It has not yet been determined if ectopic decidua and endometriosis are simply different forms of the same phenomenon. The morphology of ectopic endometrial tissue varies greatly, ranging from simple islands of more or less undifferentiated cells to large areas of highly differentiated endometrium. These tissue areas react in varying degrees to ovarian hormones (Schweppe et al. 1984; Schweppe and Wynn 1984; Schmidt 1985). Unlike in eutopic endometrium, the level of binding sites for estrogen remains constant throughout the cycle (Gould et al. 1983). The number of hormone receptors is also much lower in ectopic than in eutopic endometrium (Vierikko et al. 1985).

1.7 Basis for This Study

The introduction of in vitro fertilization (IVF) refocused interest on the parameters of implantation (Edwards and Purdy 1982). Over 100 embryos had to be transferred before the first embryo was implanted successfully (Jones 1982), but the IVF rate of implantation was soon increased to about 20% – less than half of the estimated 45% under physiological conditions (Edwards and Fishel 1982). Among the parameters important for a successful initiation of pregnancy are the type of cervical mucus [E_s, E_l, or G] (Höglund and Odeblad 1977), sperm capacitation (Zanefeld 1978), and the condition of the endometrium (Dallenbach-Hellweg 1984). In the endometrium itself, the endometrial enzymes play an important role (Wilborn and Flowers 1979), as do the cellular elements of the stroma particularly the endometrial granulocytes (K cells) and the predecidual cells (Dallenbach-Hellweg 1975), which should be optimally differentiated for implantation. After ovulation the glandular epithelial cells of the endometrium develop a prominent NCS which is not formed in anovulatory cycles (Spornitz, unpublished observation). The NCS is possibly not only an indicator of ovulation, but also a mediator of differentiation in the endometrium; the formation of phosphorylase and acid phosphatase might be initiated through the NCS (Wynn 1978).

The endometrium does not only play an outstanding role during implantation but is also a major target for methods of contraception. Two kinds of contraception in particular are dependent upon the participation of the endometrium: intrauterine contraception (cf. Hefnawi and Segal 1975; Beller et al. 1984), and hormonal contraception, i.e., in cases where the contraceptive does not suppress ovulation, such as luteal supplementation (Minipill) (Mall-Haefeli et al. 1976). Various mechanisms have been postulated to explain the mode of action of intrauterine contraceptives.

Inert IUDs are believed to exert their contraceptive effect mainly through an FBR (Elstein and Ezenyirioha 1984). Copper IUDs are believed to induce an even higher infiltration of leukocytes than inert IUDs, which is responsible for the elevated rate of prostaglandin synthesis, which in turn causes the uterine contractions that make implantation impossible (El Badrawi and Hafez 1980). Moreover, the intrauterine milieu is also altered as a result of the copper released from the IUD. These changes probably include the inactivation of enzymes, and toxic effects on the blastocyst and ascending spermatozoa (Wagner 1976). The asynchronous differentiation of the uterine tissue, caused by the copper released from copper IUDs is also believed to be part of their contraceptive effect (Wynn 1968; Ludwig and Metzger 1984). The progesterone IUD induces endometrial decidualization which is followed by glandular atrophy (Spornitz et al. 1983), both of which probably form part of the contraceptive action of the progesterone IUD.

During the postovulatory phase of the physiological, i.e., untreated cycle, ultrastructural changes have been reported which are collectively called the postovulatory triad (Cornillie et al. 1985). The postovulatory triad comprises the NCS, subnuclear glycogen, and giant mitochondria. The postovulatory triad appears to be directly related to the postovulatory hormonal changes, particularly the surge of progesterone. There is increasing evidence that the postovulatory triad plays an important role in contraception as well as in the events leading up to implantation. It is the aim of the present study to investigate the exact ultrastructure of the postovulatory triad, including the initial steps in glycogen formation; the formation, morphology, and regression of the

NCS; and the three-dimensional conformation of giant mitochondria – to name only a few.

Particular attention will be paid to the mode of formation, mature morphology, and degradation of the different types of decidual cells, as well as to their participation in implantation and contraception. Furthermore it is the aim of this study to clarify the possible origin of K cells (endometrial granulocytes), their mode of formation, ultrastructural characteristics, and ultimate fate. K cells are believed to play an important role in implantation (Dallenbach-Hellweg 1975).

Under physiological conditions as well as for IVF or gamete intrafallopian transfer (GIFT) the maintenance of optimal tissue parameters seems to be the sine qua non for a successful implantation. Nevertheless, ectopic pregnancies do occur where the blastocysts implant in tissues like the mesentery, ovary, or oviduct. The ectopic tissues and their state of differentiation are by no means comparable to an optimally differentiated endometrium. For this reason, tubal pregnancies will be investigated in this study in order to elucidate the presumed participation of endometriotic tissue in the establishment of ectopic pregnancies.

2 Material and Methods

2.1 Material

In the present study, endometrium from 157 women was investigated. More than half of the tissue studied was taken from hysterectomized uteri; the rest was obtained through curettage. Since the use of hysterectomy tissue permits better localization within the uterine cavity, primarily this type of tissue was investigated. Biopsy tissue was only secondarily taken into account, particularly in cases where the exact localization within the uterine cavity was not so important. It was demonstrated that electron microscopic and histologic findings differ greatly according to the localization of the tissue within the uterine cavity. This was particularly true of endometrial tissue taken from patients with IUDs, where a direct relationship was found between the distance of the biopsy from the IUD and changes in tissue morphology. The tissue investigated was obtained from:

- 41 Patients with progesterone IUDs.
- 67 Patients with copper IUDs.
- 24 Patients without treatment.
- 25 Patients with pregnancies.

Generally the tissue was taken at various points of the cycle, i.e., between day 4 and day 32. The exact day of the cycle was determined on the basis of histologic sections according to the criteria of Noyes et al. (1950) and Dallenbach-Hellweg (1975). In cases where dating was not possible because of the drastic changes induced by IUDs etc., the day of the cycle was determined on the basis of the date of the last menstruation as given by the patients.

For 15 patients with progesterone IUDs a hormonal profile was established which served as a basis for the determination of the exact day of the cycle.

The pregnant uteri were either obtained from patients who underwent abortion for medical reasons between week 6 and week 12 of pregnancy, or they were obtained during the last trimester of pregnancy, i.e., after birth.

2.2 Light Microscopy

For light microscopic investigations the tissue was fixed in Carnoy for between 1 and 24 h (depending upon the size of tissue block and whether obtained by curettage or hysterectomy). Fixation was followed by dehydration in ascending series of alcohol. After intermediate treatment in butanol, the tissue was then embedded in Paraplast (Athy, Kildare, Ireland). Tissues were investigated using the "five-sections technique" which employs five different stains to examine five consecutive sections. The stains used were: (a) azan, (b) thionin, (c) Gomori (silver impregnation), (d) periodic acid-Schiff (PAS) and (e) Staining for glycogen (reaction according to Bauer). The staining formulas and procedures were taken from Romeis (1968).

2.3 Transmission Electron Microscopy

For electron microscopic investigations, the tissue must be fixed immediately after removal because of the rapid onset of autolysis. Therefore, the tissue was already cut to appropriate size and immersed in the fixing solutions in the surgery or operating theater.

Hysterectomized uteri were cut open along the lateral aspects of the corpus uteri, beginning at the cervix. The endometrium was carefully mapped and tissue was taken from well-localized areas, such as the uterotubal junction, upper and lower corpus, anterior and posterior wall. Tissues from different areas were all processed separately to assure positive identification throughout the investigations.

When changes to the uterine epithelium were of a general nature, biopsy tissue could always be easily interpreted, but when the changes were local – such as those caused by IUDs – tissue removed by curettage was of little value.

Two methods were employed for tissue fixation:

1. single fixation with osmium tetroxide (OsO_4).
2. double fixation with glutaraldehyde (GA) followed by OsO_4.

GA (2%) was added to a phosphate buffer (Millonig) of pH 7.6 and 300 mosmol, resulting in a total osmolarity of 500 mosmol. OsO_4 1% was also added to Millonig's buffer, making a total osmolarity of 330 mosmol. After fixation the tissue was rinsed in Millonig's buffer, dehydrated in acetone, and embedded either in Araldite (Ciba-Geigy, Basel, Switzerland) or Epon 812 (Fluka, Buchs, Switzerland). Fixation and dehydration was achieved at 0°–4°C. Impregnation and embedding was performed at room temperature. To assure reproducible cutting quality of tissue blocks, embedding was usually done on the same day that the tissue was removed.

Semi-thin sections (1–1.5 µm) and ultra-thin sections (approx. 60 nm) were cut on Reichert microtomes (OmU3 and Ultracut E). Semi-thin sections were stained with p-phenylenediamine; ultra-thin sections were stained with uranyl acetate and lead citrate. All sections were examined with either a Philips 301, or a Zeiss 10 C electron microscope.

2.4 Scanning Electron Microscopy

For scanning electron microscopy (SEM) two kinds of specimen were investigated:

1. Specimens prepared directly for SEM.
2. De-embedded transmission electron microscopy (TEM) blocks.

Tissue prepared for SEM was fixed in the same way as for TEM – with either single OsO_4 fixation or double fixation with GA followed by OsO_4. The specimens were rinsed in buffer, dehydrated in acetone, and critical point dried in CO_2 (Cohen 1979; Ludwig et al. 1976a). Critical point drying was done in a Balzer's unit. Dried specimen were then mounted on aluminum stubs and sputter coated with 15 nm of gold, at 4°C. In most cases SEM immediately followed preparation; where this was not possible, specimen were stored under atmospheric pressure in a desiccator which contained a desiccating agent (Silikagel). De-embedding was generally done with tissue blocks, from which ultra-thin sections had already been cut, using a modification of the method described by Mayor et al. (1961). The tissue blocks were freed mechanically (razor blades) from excessive epoxy resin, incubated in a 3.3% sodium methylate solution which had been dissolved in a mixture of methanol/toluol (2 : 1), and put on a shaker for 15–120 min. They were then washed in a methanol/toluol solution, dehydrated in acetone, critical point dried, and sputter coated with 15 nm gold.

2.5 Three-Dimensional Reconstruction

For three-dimensional reconstructions, two different methods were employed:

1. Manual reconstruction.
2. Computer-aided reconstruction.

Manual Reconstruction. For manual reconstruction the organelles were cut by ultra-thin serial sectioning. In order to cover the full extension of the NCS and giant mitochondria, the minimal number of sections necessary was first calculated, and then at least twice as many sections were cut. The sections were collected on collodion-covered slot grid. Based on the thickness of the sections, the dimension of the organelles, and a security factor calculated into it, up to 40 grids per single organelle had to be investigated. Generally photography started in the middle of a series, to assure full coverage of the selected organelle. Series with missing or low quality sections were dismissed. From every photographic negative enlarged drawings were produced on styrofoam plates. Benchmarks were added and the contours of the organelles drawn on the styrofoam were cut out with a heated saw. The thickness of the plates was calculated according to the thickness of the original ultra-thin section and the total enlargement factor. Styrofoam plates are available in a full range of thicknesses from 2 mm to several centimeters. With the aid of the benchmarks, the styrofoam plates were mounted on top of each other and a cast of the resulting negative mold was made with Araldite (Araldite SV 426 and Hardener HV 426, Ciba-Geigy, Basel, Switzerland). This type of epoxy resin is particularly well suited to being worked on with finishing tools. The resulting casts were a perfect likeness of the original organelles, except for those areas which had been tangentially sectioned – particularly where the original ultra-thin section was the same size as the ultrastructural detail cut from it. In order to achieve an optimal resolution, the thickness of the sections was kept around 40 nm; nevertheless, with tangentially sectioned details, a steplike contour to the cast was inevitable. These steps were leveled out with epoxy resin.

Computer-Aided Reconstruction. Serial sections were cut in the same way for computer-aided reconstruction as for manual reconstruction. From the enlarged photographic prints, contours of the organelles were drawn on cellophane foil and then digitalized on a computer with a video camera. To computing of the three-dimensional structure of the digitalized images, volume elements (or voxels) were used (Ylä-Jääski and Kübler 1987a). In a voxel computation, the data amount and speed is independent of the object complexity, whereas in vector-based computations up to millions of vectors are required for an accurate representation. For the actual conversion of the data into a binary three-dimensional array, an automatic command procedure was run in a DeAnza IP 8500 image processing system. This procedure consists of binarization, filling the contours, and interpolation until an isotropic three-dimensional resolution is achieved. The system comprises two fast microprocessors, operating in parallel mode. Each processor has a fast direct memory access to 8 Mbytes of data memory for storing the voxel data. In contrast to vector-based display systems the direct display from voxels requires considerably more data memory and is usually also extremely slow. Therefore, a new reconstruction algorithm was used which made it possible to generate an oblique view from a data matrix of 256^3 voxels in about 6 s (Klein and Kübler 1986, Ylä-Jääski and Kübler 1987b).

2.6 Histochemistry

To determine the content of secretory granules, decidual granules, and their precursors, we observed enzymatic digestion on ultra-thin sections. Three different enzymes were used (Monneron and Bernhard 1966):

1. Pronase, as a broad-spectrum proteolytic enzyme.
2. α-amylase, for the detection of glycogen.
3. Pancreas lipase, for the detection of lipid constituents.

Differentiation between glycogen and ribonucleic acid (RNA) particles (in areas of glycogen deposits) was achieved with differential staining according to Bernhard (1968). For selective staining of glycogen particles silver staining according to Thiéry (1967) was used.

2.7 Morphometry

For morphometric measurements the point counting method was employed (Weibel et al. 1966; Weibel 1969). Morphometry was used to determine the volume fraction of various cell components like ER, Golgi apparatus, mitochondria, paraplasma, nucleus, and decidual granules. Point counting was done on electron micrographs at two different levels, i.e., with magnifications of 12 000 and 37 500. A total of 160 cells were taken at random and studied; some of them were so large that up to four photographs of each was needed. In total, over 300 photographs were used to determine the volume fractions.

Acknowledgements. Most of the material investigated was obtained from the Frauenspital der Universität Basel, Switzerland. Several uteri and biopsies were also obtained from: Bezirksspital Breitenbach, SO, Switzerland; Bezirksspital Zofingen, SO, Switzerland; Universitäts-Frauenklinik Bern, Switzerland; and the Department of Obstetrics and Gynecology, Wayne State University Detroit, MI, USA.

3 Results

3.1 Physiological Postovulatory Endometrium: Glandular Epithelium

3.1.1 The Nucleolar Channel System

3.1.1.1 Problems with Three-Dimensional Reconstruction

A dynamic entity as complex as the NCS can hardly be understood and interpreted on the basis of serial sections alone. Moreover, the fact that the NCS develops from the nuclear envelope within hours or days, appears to be stable for only a few days and disappears from the cells before the onset of desquamation makes the task even more difficult. During formation and involution the morphology of the NCS changes constantly. Thus, any attempt to visualize the exact morphology of the NCS is bound to fail. For this reason, a large series of serial sections was made from NCS at the height of their formation, i.e., around days 19 and 20 of the female cycle. These series were then used in manual and computer-aided reconstruction. A section through the widest part of the NCS reveals that it consists of up to 54 individual tubules. The computer program available for reconstruction was not able to handle such a vast amount of data for up to 30 sections through an individual NCS. Therefore, computer-aided reconstruction utilized only up to one-third of the NCS and generally at least three parts of the reconstruction had, thus, to be fitted together manually. Consequently the old-fashioned wax-plate reconstruction method was employed in a modernized version. Styrofoam plates were substituted for wax plates and Araldite for plaster as described in the previous chapter.

3.1.1.2 The Reconstructed NCS

Both manual and computer-aided reconstruction showed that the NCS has a polar organization, with one pole attached to the nuclear membrane (Fig. 1) and the other facing the nucleoplasm (Fig. 2). The morphology of the pole attached to the nuclear membrane varies considerably, depending upon the day of the cycle. Reconstruction which did not include the nuclear membrane shows that the NCS has an opening in the attachment area which goes into its core. This opening is about 400 nm wide, with a circumference of about 1250 nm. In the attachment region it can be seen that up to four individual NCS tubules are connected to the nuclear membrane; the intratubular space is,

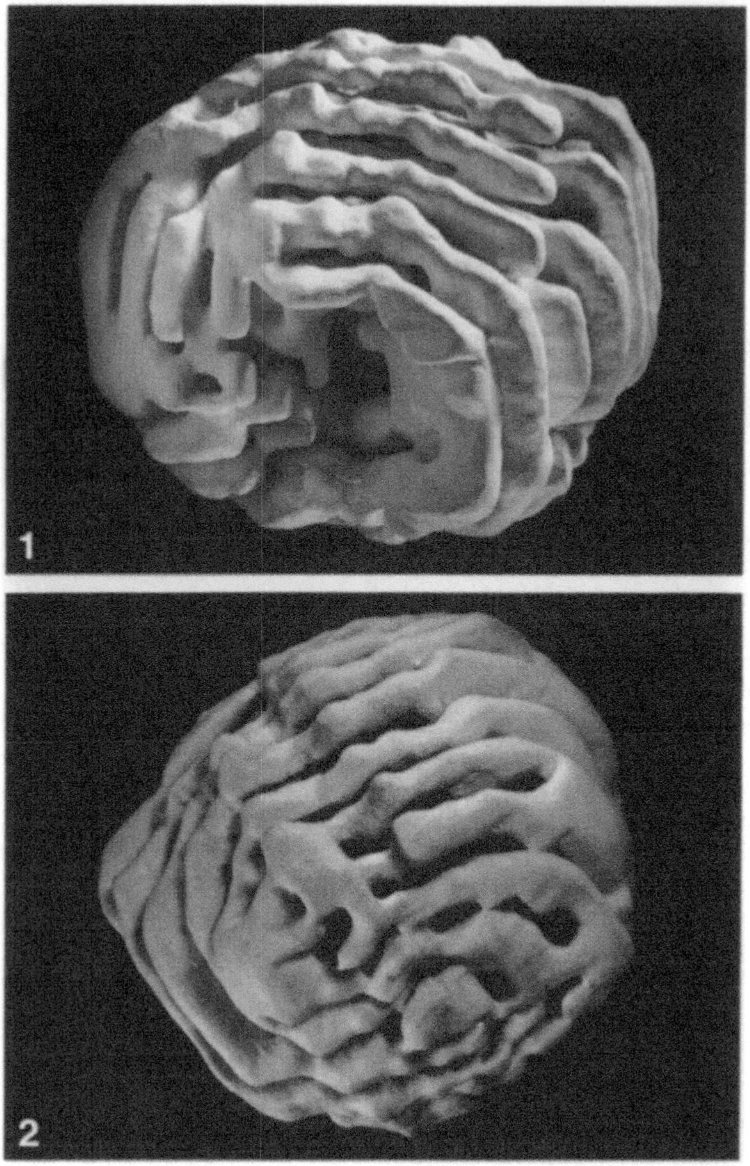

Figs. 1, 2. Manually reconstructed NCS. **Fig. 1.** Site of nuclear membrane which opens into core of NCS. On *left*, there are four tubules which were connected to the nuclear membrane. **Fig. 2.** Karyoplasma pole is shown with various interconnecting tubules

thus, a continuation of the perinuclear space. At the attachment site the tubules are generally somewhat constricted. For this reason it is not always possible to detect a continuity between the perinuclear and the intratubular compartments, particular in cases where this continuity is only tangentially sectioned. Serial sectioning, however, always indicates whether continuities are present. Depending upon the day of the cycle there may be fewer continuities,

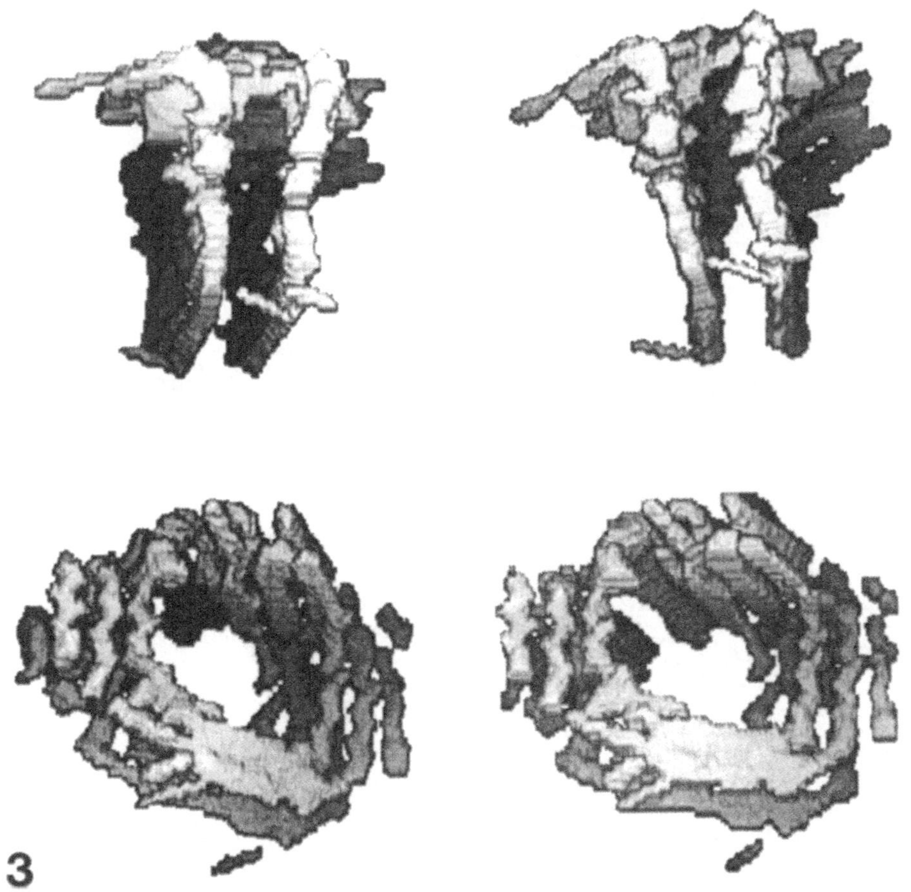
3

Fig. 3. Computer-aided reconstruction of part of NCS. The four pictures were taken from the same NCS but with different angles of inclination. Ruffles on tubules of *upper left picture* correspond to half the thickness of a single section. They are caused by the calculated interpolation between individual sections

and they do not always seem to be patent. Membrane continuities are, however, always present. At the height of their existence, i.e., around days 19 and 20, there are always four attachment sites present and from these four points seven NCS tubules are formed through branching. The individual tubules measure about 100–120 nm in diameter and run spirally around the core of the NCS. Relative to the longitudinal axis of the NCS, the angle of inclination of the spiral is 25°. From the site of attachment at the nuclear membrane through to the nucleoplasm pole the spiral runs in one and a half turns. Thus, the individual tubule can be seen twice when viewed from the side. In places where the tubules run through the equator of the NCS, they can only be seen once when viewed from the side. Most of the tubules are confluent in the area of the nucleoplasm pole where they form a common "polar plate" (Fig. 2). From this plate further tubules run into the core of the NCS constituting a second and a third layer of tubules which runs parallel to the outer layer of tubules (Fig. 3).

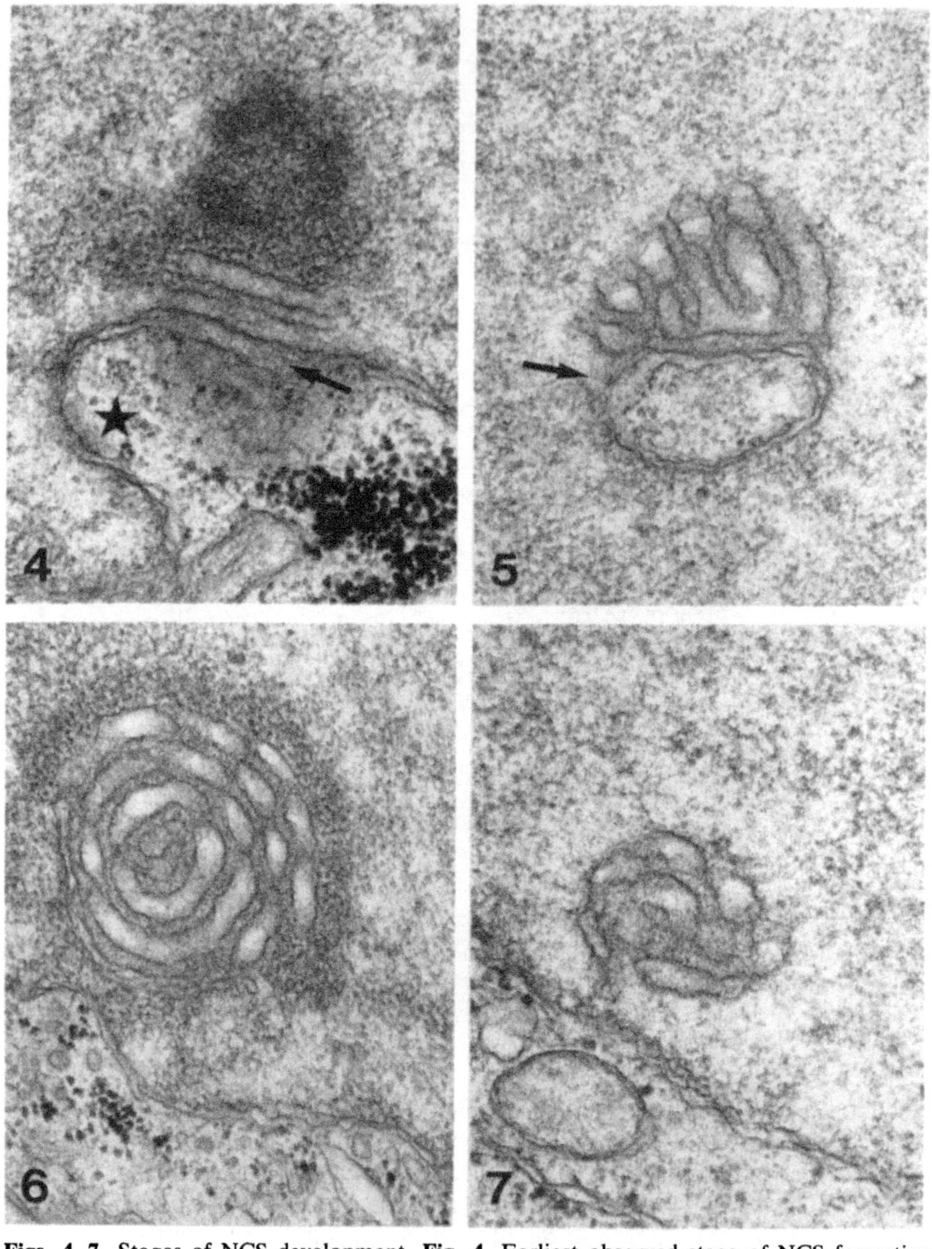

Figs. 4–7. Stages of NCS development. **Fig. 4.** Earliest observed stage of NCS formation (day 14). The nuclear membrane is slightly dilated (*arrow*). A second tubule has been formed already. The nuclear membrane has started to invaginate (*asterisk*), ×46000. **Fig. 5.** The invagination of the nuclear membrane is cut transversely. The dilation of the perinuclear space (*arrow*) is typical for the NCS formation site (day 15). The cytoplasmic side of the nuclear membrane is studded with ribosomes, ×42000. **Fig. 6.** An early stage of tubule coiling. Numerous ribosomes are present in the surrounding matrix (day 15), ×42000. **Fig. 7.** An early coiling tubule which is connected to the nuclear membrane through a constricted tubular area (day 15), ×53000

The middle and inner layer of tubules also runs spirally towards the pole attachment area at the nuclear membrane. Computer-aided reconstruction clearly shows interconnections between the individual tubules, i.e., between tubules from the same layers as well as between tubules from different layers (Fig. 3). Because of these interconnections cisternae are frequently formed which may extend from the outer to the inner layer of tubules. Supercoiling of individual tubules was, however, not found with either manual or computer-aided reconstruction.

3.1.1.3 The NCS on Sections

3.1.1.3.1 Formation of the NCS

The initial stages of NCS formation are difficult to detect on sections (Figs. 4–7), since they look like simple dilatations of the nuclear membrane. The fact that they are only short-lived might also explain why the initial stages have so far not been described. They are, nevertheless, easy to distinguish from terminal stages of the NCS owing to the matrix condensations which are only found in the latter stages. The first signs of NCS formation can be seen in those areas where nucleolar heterochromatin is attached to the nuclear membrane; in these areas the perinuclear space becomes dilated (Fig. 4). The inner membrane protrudes into the nucleoplasm and, in time, the protrusions become extended into tubular structures. At the site of attachment to the nuclear membrane a constriction appears which temporarily blocks the passage between the perinuclear space and the intratubular lumen (Fig. 7). Subsequently, the tubular extensions elongate, branch, and coil up to form a primitive NCS which may already consist of up to the three rows of tubules, such as are typical for the mature NCS (Fig. 8). An invagination of the nuclear membrane also accomparies the first steps of NCS formation, sometimes penetrating the nucleoplasm to about one-third of the nuclear diameter (Fig. 9). Thus, one side of the NCS is located in the middle of the nucleoplasm while the other is directly attached to the nuclear membrane, thus establishing a possible pathway into the cytoplasm. With further maturation the attachment to the nuclear membrane is either obstructed through constrictions or entirely obliterated.

3.1.1.3.2 The Mature NCS

Since the dynamics of the NCS show it to be a rapidly changing entity, we have only investigated sections from glandular epithelial cells from days 19 and 20 which show the NCS at the height of its development. Sections through the central region of the NCS are the most distinct and easiest to interpret, particularly when the tubules are sectioned exactly transversely. Such a section reveals that the NCS comprises three rows or layers of tubules which are arranged around the core in a circular manner. The innermost layer, therefore, has the smallest diameter and consists of fewer individual tubules than the outermost or middle layer. The highest number of individual tubules found on an ideal transverse section are: 22 in the outer layer; 20 in the middle layer; 14 in the inner layer. Thus, the maximum number of tubular cross sections found

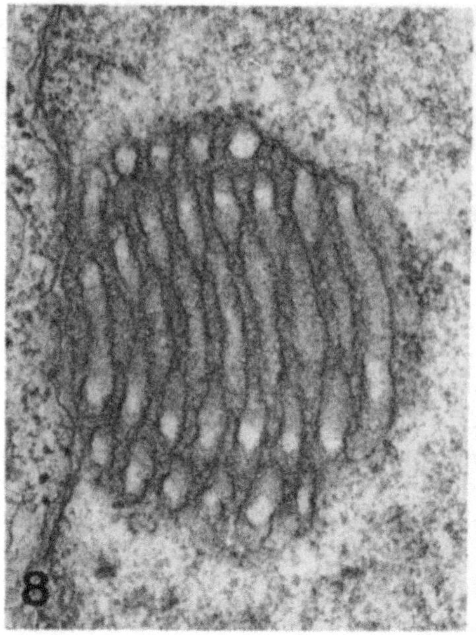

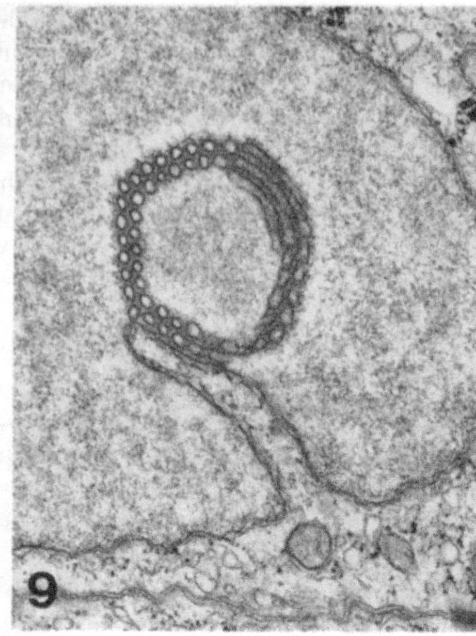

Fig. 8. NCS section at right angles to the tubules. The fact that this section was made through the peripheral zone of the NCS makes it possible to see three rows of tubules (day 18), ×46000

Fig. 9. A deep invagination of the nuclear membrane. Because of the high angle of inclination at which the tubules are turning it is possible to see both tangentially and cross-sectioned ones (day 17), ×20000

is 56. On sections closer to the polar regions the number of tubular cross sections found is much lower, largely due to the fact that in these areas the tubules are frequently sectioned in a tangential manner.

In the area of the opening of the NCS, i.e., at the attachment site, there are no tubules. Measured in relation to the outer circumference of the NCS this area represents about 25°. Serial sections demonstate that sections through polar regions result more and more in tangential cross sections. They also show that the angle of inclination between tubules from opposite poles amounts to 120°, which is a good indication of the spiral course of the tubules (Fig. 10). The individual tubules are embedded in an electron dense material which is apparently, to a large extent, made up of ribosomes (Fig. 10a–1).

3.1.1.3.3 Ultimate Fate of the NCS

At the time of maximal development of the NCS, there are very few electron dense particles in the core region (Fig. 10e), a situation which changes as soon as involution of the NCS starts. The first sign of regression of the NCS is the appearance of an electron dense matrix in the core region, such as is found in intact mitochondria. Concomitantly the tubules start to lose their high degree of order. A process of segmentation, begins which results in vesicular fragmentation of the tubules. Fragmentation sometimes reaches a level where most of the tubules are made up of vesicles alone, embedded in an electron dense

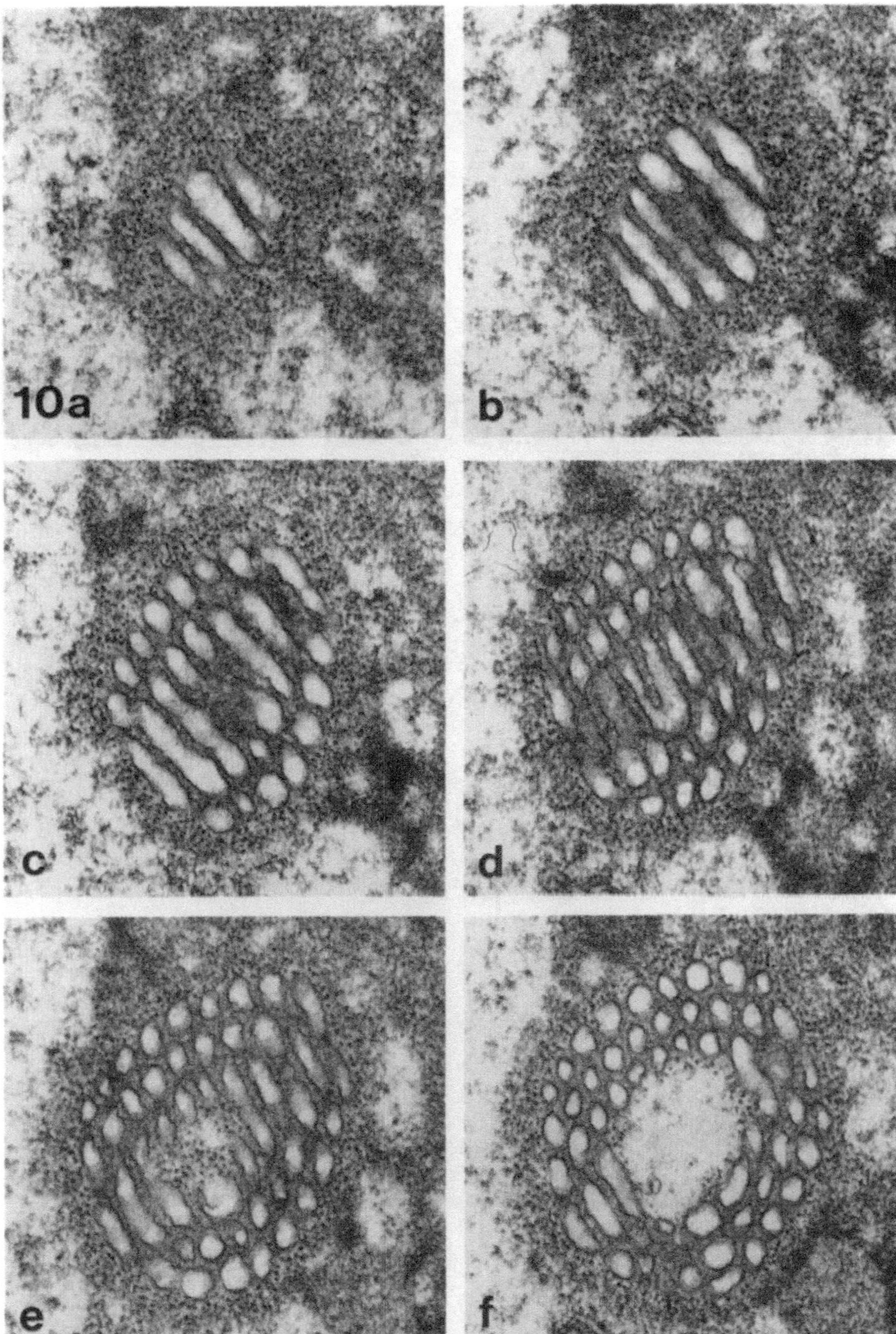

Fig. 10 a–1. Series of sections through a fully developed NCS from day 19 of the cycle. Every second section of the series is shown. The whole series consists of 25 individual sections. Between section **a** and section **1** the tubules have turned by 90°, ×46 000

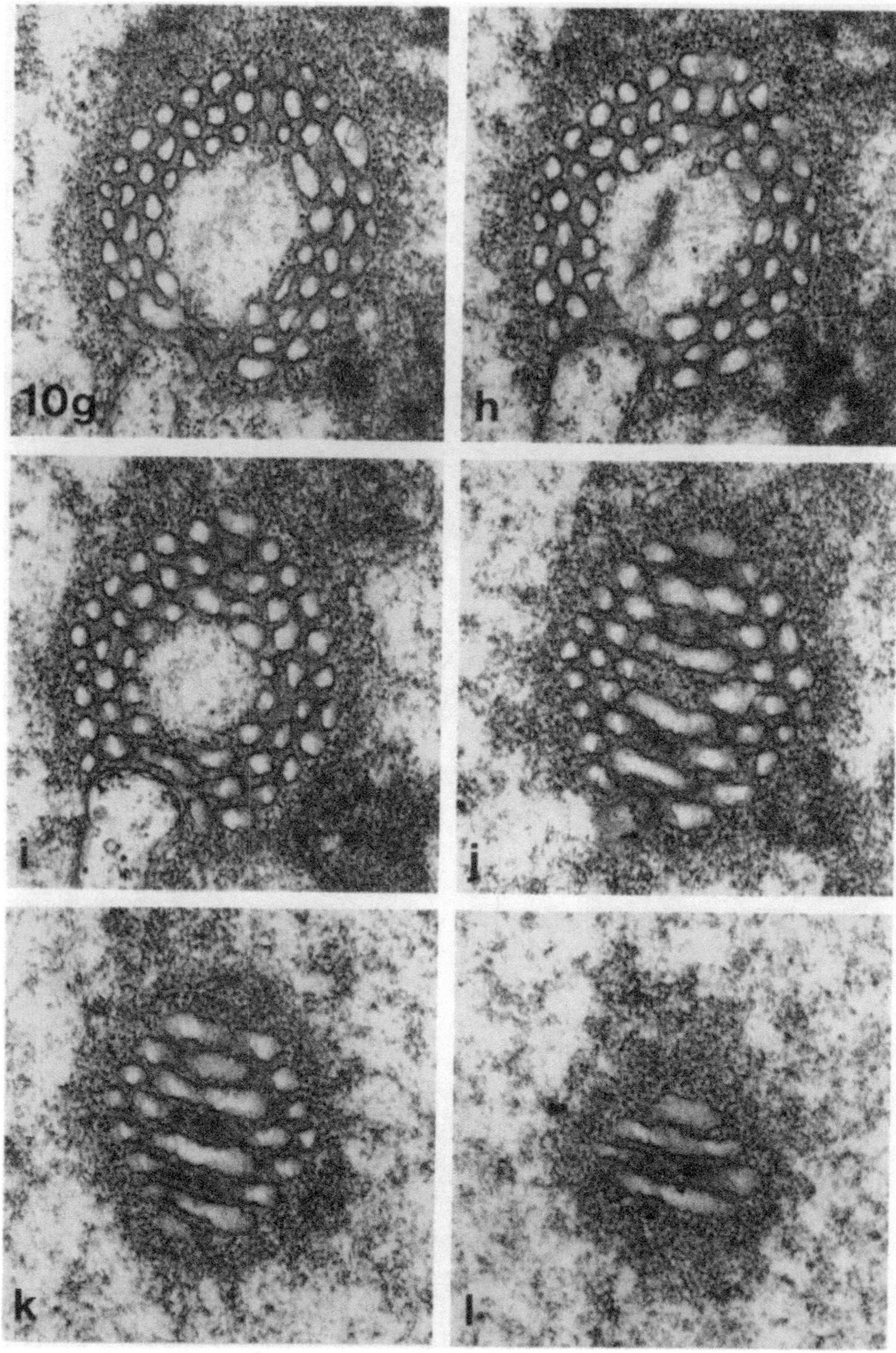

Fig. 10. (Continued)

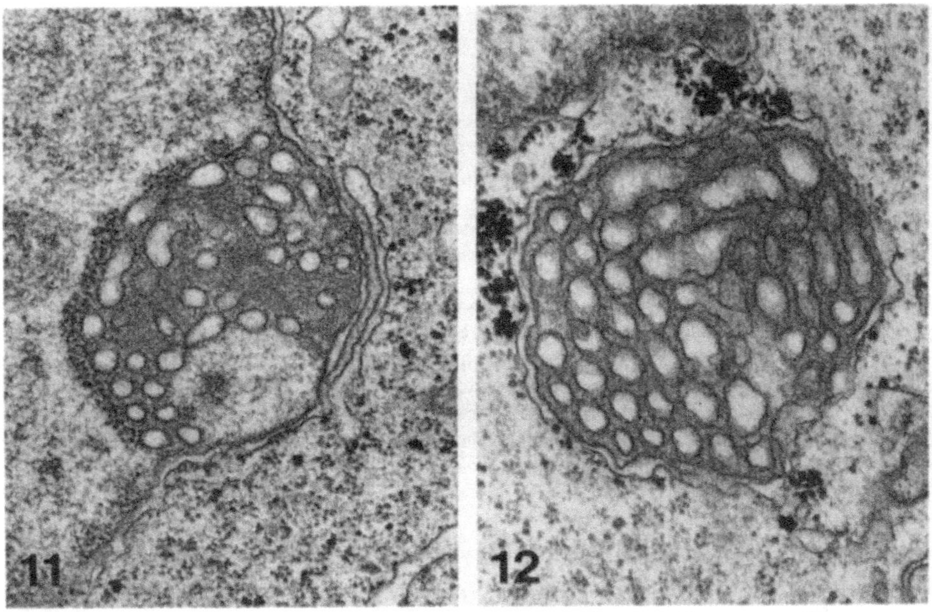

Fig. 11. Nuclear membrane bulges into the cytoplasm just before extrusion of NCS (day 25), ×37 000

Fig. 12. NCS has been expelled from the nucleus (day 26). On both figures the NCS tubules have fragmented into vesicles, ×37 000

matrix (Fig. 11). This electron dense matrix fills the core region and the immediate surrounding of the NCS and consists to a large extent of ribosomes. The fragmenting NCS loses its contact with the nucleolus, the number of tubular profiles is drastically reduced, and the NCS moves close to the nuclear membrane. Around day 24 the NCS pushes the nuclear membrane into the cytoplasm (Fig. 11); frequently the NCS can be found to extrude into the cytoplasm where it lies within a vesicular bud of the nuclear membrane (Fig. 12). Serial sections show that the NCS vesicles are not connected to the nuclear membrane surrounding the nucleus; they are located completely outside the nucleus. Occasionally, as early as day 22, autophagic vacuoles can be found in the glandular epithelial cells, containing among other cellular remnants, fragments of the NCS. Generally the NCS has completely disappeared from the cells around day 28. It is unusual to find NCS remnants in the cells shortly before desquamation, and when found they are still connected to the nuclear membrane – this might explain why they were not subject to intracellular digestion.

3.1.2 Giant Mitochondria

3.1.2.1 Problems of Reconstruction

In the computer-aided reconstruction of giant mitochondria the same procedure was used as for the reconstruction of the NCS. Mitochondria of the

Table 1. Stages in Formation and Regression of the NCS

Day	Feature
14	Initial stages of NCS formation: one to two invaginations of the inner nuclear membrane, running parallel to the nuclear membrane
15	Invaginations are transformed into tubules
16	Tubules begin to coil
17	Fully coiled NCS present, made up of two rows of tubules
18	Three rows of tubules are present
19	NCS reaches largest diameter
22	Regression starts: the core becomes electron dense
23	Distinct features of degeneration are present: tubules start to break up into fragments, the NCS becomes disconnected from the nucleolus
24	First bulges of NCS into the cytoplasm are visible
25	Autophagic vacuoles containing NCS fragments are present
28	NCS has practically disappeared from cells

middle to late secretory phase of the cycle are frequently very closely adjoined. For this reason, and because of the large number of mitochondrial protrusions, the thickness of serial sections used for reconstruction was kept well below 50 nm, to ensure that contact areas between individual mitochondria were not suppressed by the reconstruction algorithm.

3.1.2.2 Three-Dimensional Structure

The morphology of the reconstructed giant mitochondria is strictly dependent upon the day of the cycle. Around day 15 the majority of mitochondria are spherical to slightly oval in shape. They are similar in size to the regular preovulatory mitochondria. Their surface is smooth with only minute irregularities or undulations. With the aid of computer reconstruction any viewing angle can be achieved, dispensing with the need for detailed analysis of mitochondrial irregularities. However, in spite of the interpolation of individual sections, which forms the basis of computer reconstruction, irregularities between individual sections are still apparent on the final reconstructed model. These result from the varying compression of and subsequent flattening of the individual sections on the surface of the water-filled trough of the diamond knife on which the sections float after being cut (Fig. 13). From day 15 through day 20 the mitochondria grow considerably. They do, however, keep their spindle-like shape. After day 20 of the cycle they hardly show any further growth, but start developing individual protrusions (Figs. 14 and 19), which leads to their development into complex mitochondria. Generally, the mitochondria have just one protrusion or extension; however, two or three extensions are occasionally found. These extensions definitely do not make any contact with other mitochondria, particularly in the sense of connecting individual mitochondria; they are part of a reduction or fragmentation mechanism, whereby the mitochondria reduce their size prior to desquamation. The extensions reach significant dimensions, and are sometimes much longer than

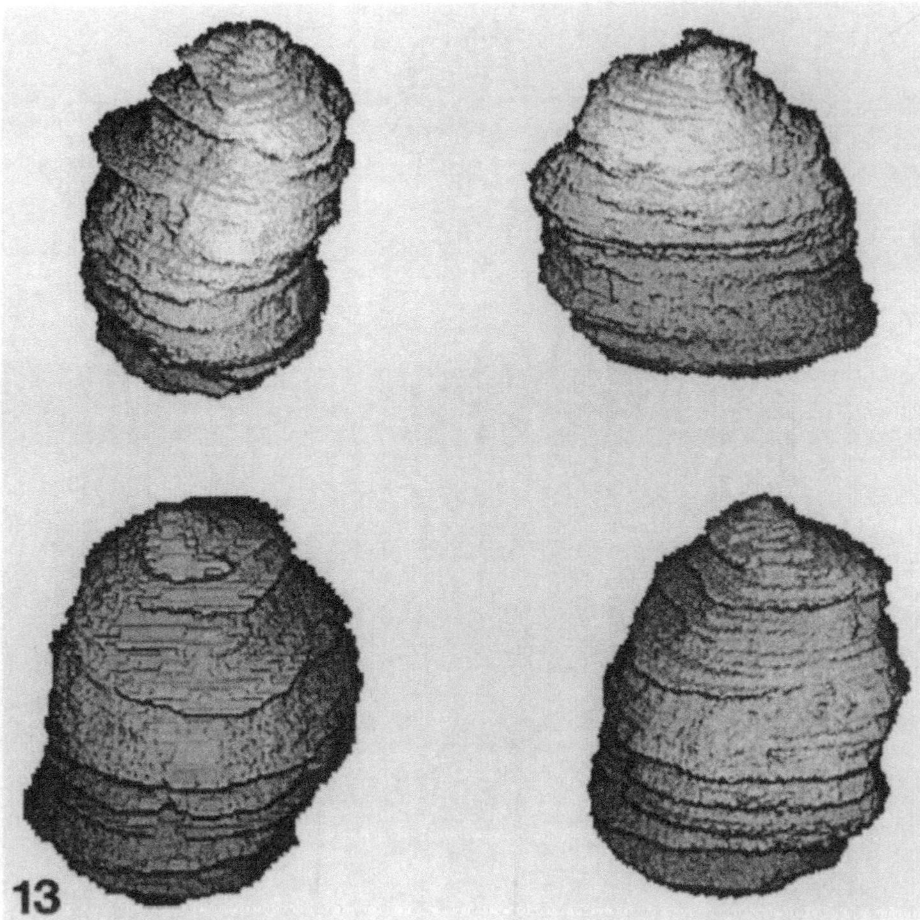
13

Fig. 13. Computer reconstruction of a giant mitochondrion from the 18th day of the cycle. Between each of the four pictures the computer image has been rotated by 90°. The circular lines on the surface of the mitochondrion correspond to half the thickness of a section. They result from the calculated interpolation between individual sections

the mitochondria themselves. Around day 24 of the cycle most of the mitochondria have reduced their size almost to preovulatory values, and around day 28 there are no more giant mitochondria present in the glandular epithelial cells. At no time does the mitochondrial population consist of only one giant mitochondrion. On the contrary, the number of individual mitochondria does not seem to be reduced in any way through the formation of giant mitochondria.

3.1.2.3 General Morphology

During the early postovulatory days many subnuclear mitochondria increase their volume. Around day 18 of the cycle only a small number are still growing,

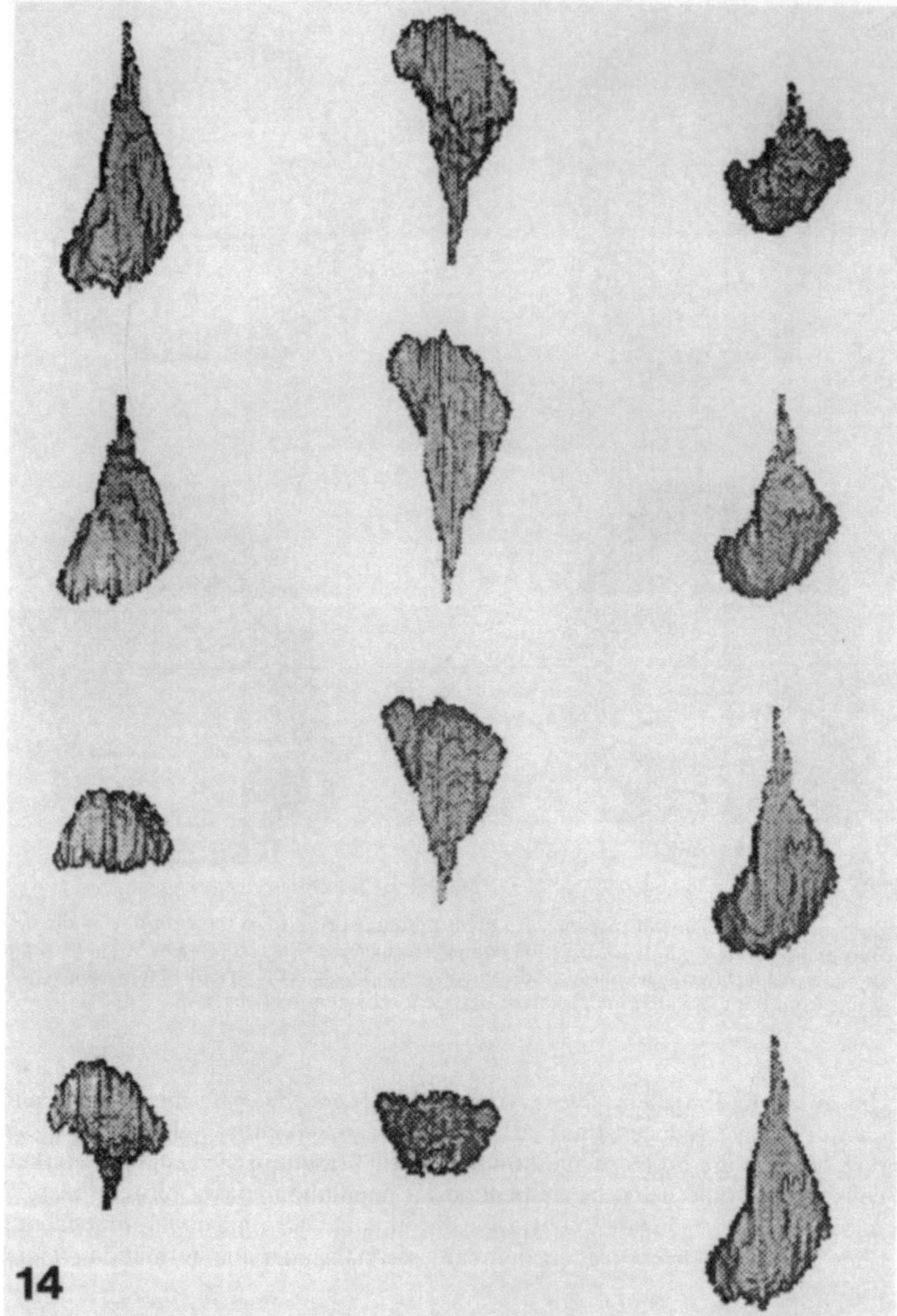

Fig. 14. A series of views of a computer-reconstructed complex giant mitochondrion from the late secretory phase of the cycle. The mitochondrion was turned exactly 30° from one picture to the next. The first of the series is in the *upper left corner*, the last in the *lower right corner*

leading to the formation of two or three giant mitochondria around days 20–22 (Figs. 15–18). The diameter of these giant mitochondria ranges between 4 and 7 µm. Occasionally they might even reach a diameter of 9 or 10 µm. In fully developed mitochondria numerous mitochondrial cristae are present which usually run perpendicular to the short axis of the mitochondria (Figs. 15 and 16). On sections parallel to the long axis of the mitochondria up to 40 individual cristae are present (Fig. 16). The distance between individual cristae lies within a range of 30–60 nm. In the matrix of routinely double-fixed giant mitochondria (GA/OsO_4), precipitates of mitochondrial DNA are frequently encountered (Fig. 17). With single fixation, using OsO_4 alone, the DNA is surrounded by an electron lucent halo (Fig. 18). With enzymatic digestion, as done with DNase on ultra-thin sections, the precipitates disappear completely. In normal (nongiant mitochondria) DNA is hardly ever encountered during the postovulatory phase of the cycle.

Giant mitochondria as well as the moderately enlarged basal mitochondria of the secretory phase of the cycle are intimately associated with RER (Figs. 16–18). Up to 95% of the mitochondrial surface is covered by cisternae of the RER which form a continuous coat, shielding the mitochondria almost entirely from direct contact with the cytoplasm. The RER membranes are separated from the mitochondria by a narrow rim of cytoplasm which is no more than 40–60 nm wide. The RER is studded on both sides with ribosomes. Mitochondrial and RER membranes were at no time seen to be connected; tangentially sectioned membranes which might sometimes be misinterpreted as a continuity were always verified as separate entities by means of a goniometer stage.

3.1.2.4 Ultimate Fate

As soon as the giant mitochondria are transformed into complex mitochondria with fingerlike protrusions, they loosen their intimate relationship with the RER; from then on they are only partially surrounded by RER (Figs. 19 and 20). Simultaneously, the rim of cytoplasm between the mitochondria and the RER widens and the RER starts to form individual smaller cisternae or even vesicles. Concomitantly the mitochondria lose their highly organized structure of parallel cristae. The cristae now begin to form a three-dimensional network (Fig. 20), which on ultra-thin sections frequently appears as darker matrix areas – due to the fact that it is more often than not tangentially sectioned. Ageing giant mitochondria are often seen in the process of transforming their cristae into vesicles. By virtue of the formation of mitochondrial protrusions which subsequently break away from the mitochondria, the volume of giant mitochondria is reduced to normal by the end of the secretory phase of the cycle. Shortly before the onset of desquamation of the endometrium, giant mitochondria have completely vanished from the cells, leaving behind only "normal" mitochondria.

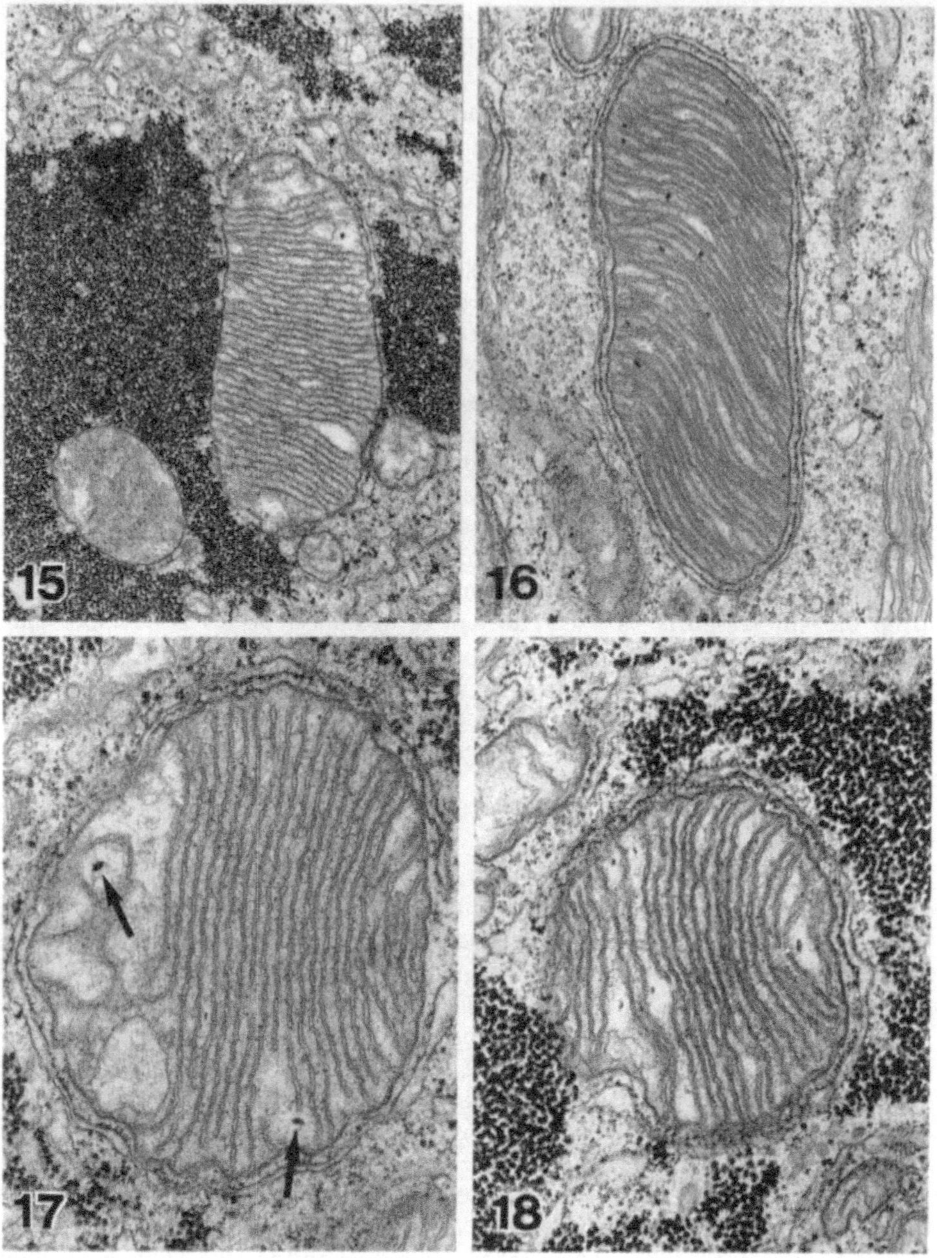

Figs. 15–18. Giant mitochondria. **Fig. 15.** An early giant mitochondrion (day 17) amidst subnuclear glycogen, ×14 500. **Fig. 16.** Early giant mitochondrion just prior to glycogen formation. The mitochondrion is entirely surrounded by RER membranes. The ribosomes on the side facing the mitochondrion are more numerous than on the cytoplasmic side, ×20 000. **Fig. 17.** Giant mitochondrion fixed in glutaraldehyde followed by OsO_4. After this fixation, precipitates are present which are probably mitochondrial DNA (*arrows*), ×31 500. **Fig. 18.** Giant mitochondrion after single OsO_4 fixation. The DNA precipitates are surrounded by an electron lucent halo, ×31 500

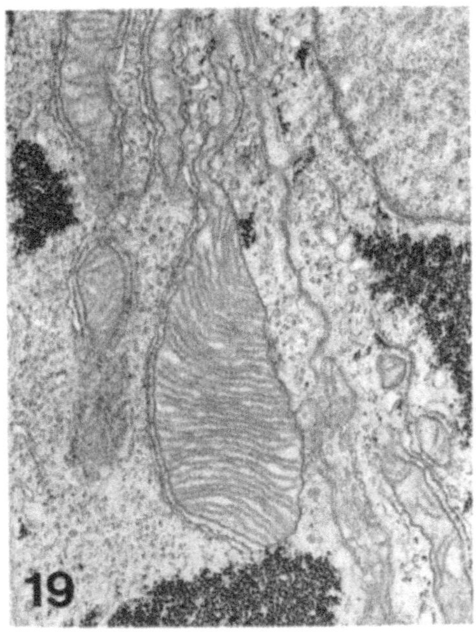

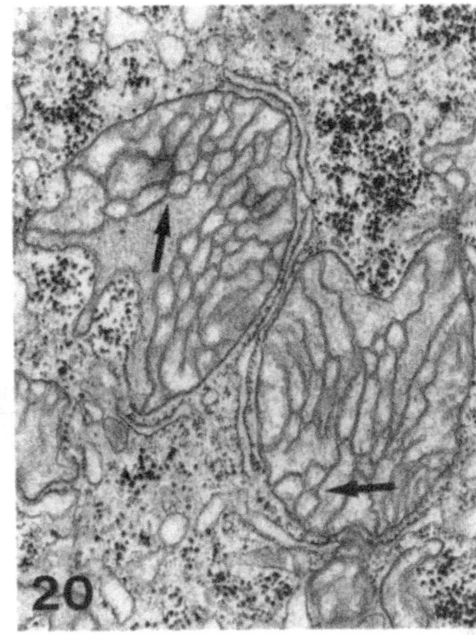

Figs. 19, 20. Giant mitochondria from the late secretory phase. **Fig. 19.** Complex mitochondrion with a long extension as they are formed towards the end of the cycle, ×14200. **Fig. 20.** Degenerating giant mitochondria with reticular mitochondrial cristae (*arrows*), ×22500

3.1.3 Glycogen in Glandular Epithelial Cells

3.1.3.1 Initial Stage of Glycogen Formation: Polysomes

Before initial stages of glycogen formation can be detected in the glandular epithelial cells, the number of subnuclear polysomes and free ribosomes is drastically multiplied. The increase occurs between days 8 and 12 of the cycle (Fig. 21). Spiral polysomes are only rarely found; the straight form predominates in the cells. The polysomes are studded on both sides with ribosomes (Fig. 21), and the space between these two rows of ribosomes, which possesses a low affinity for stains, measures about 7 nm. Short exposure to RNase renders the material between the ribosomes absolutely unstainable. The ribosomes themselves need longer exposure to become digested. Around days 14–15 of the cycle the first signs of glycogen formation can be detected in the areas of the polysomes. The first specifically stainable glycogen particles are formed on the polysomes and glycogen appears to be deposited directly onto the ribosomes (Figs. 22, 23). Using the Thiéry procedure for selective staining during the initial stages of glycogen formation (Thiéry 1967), not only free polysomes but also some of the ribosomes on the RER can be stained. Glycogen deposition is present in varying degrees, ranging from very little evidence of glycogen on the ribosomes to seemingly pure glycogen particles on the polysomes. With progression of glycogen formation, the basal or subnuclear region of the cells becomes almost entirely filled with glycogen. As a

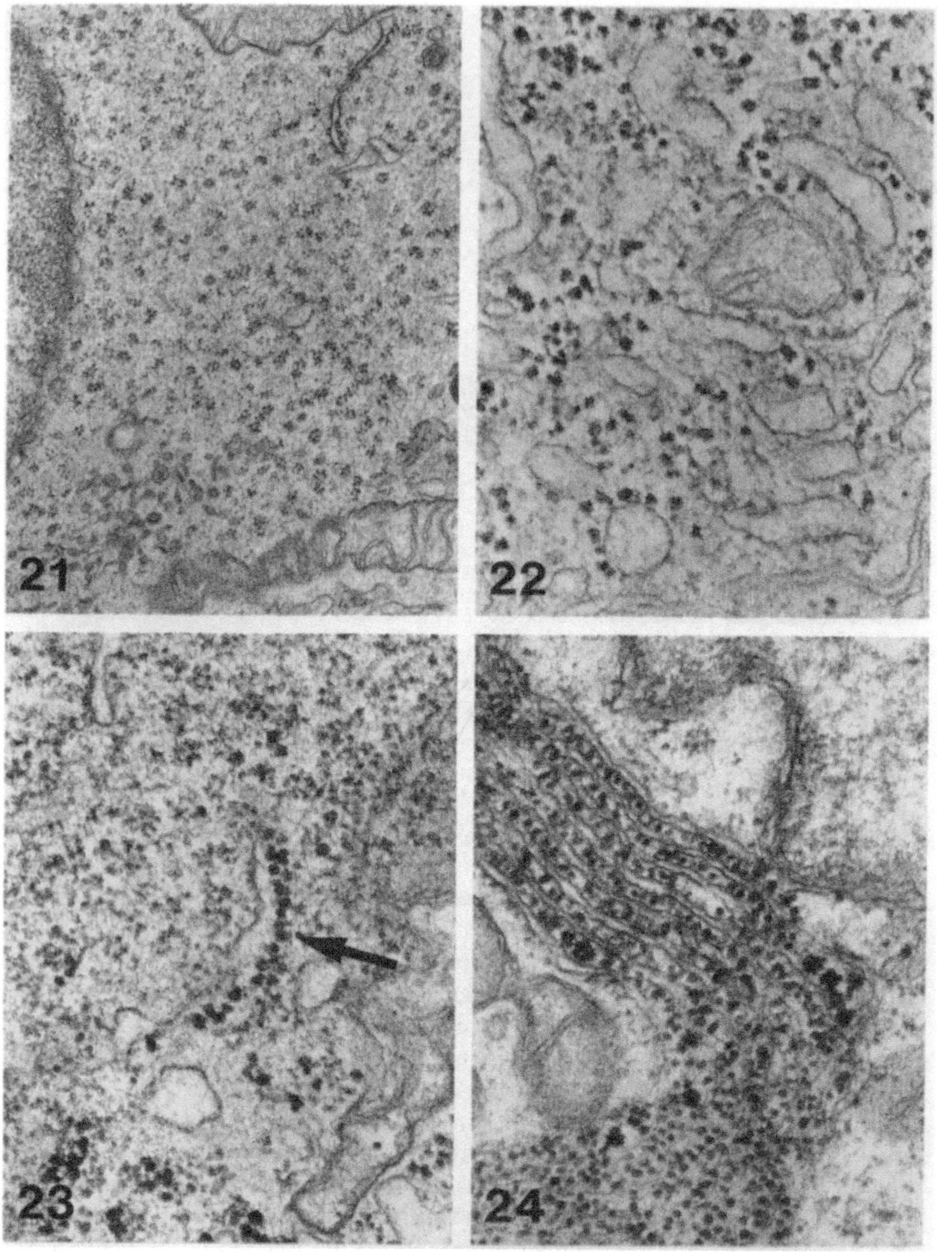

Figs. 21–24. Stages of glycogen formation. **Fig. 21.** Subnuclear region of an epithelial cell just prior to glycogen formation. Numerous polysomes studded on both sides with ribosomes are present, ×24 700. **Fig. 22.** Initial stage of glycogen formation. The RER vesicles and the polysomes are studded with both glycogen particles and ribosomes. A particular glycogen stain was used. The ribosomes (e.g., right under the mitochondrion) are unstained, ×80 000. **Fig. 23.** Polysome area from the early postovulatory period. Glycogen formation has started already. An RER cisternae is studded on one side with ribosomes and on the other with glycogen (*arrow*), ×60 000. **Fig. 24.** An RER coil as they are occasionally present in the immediate postovulatory period. This RER coil is continuous with the nuclear membrane. The smaller particles are ribosomes, the larger particles glycogen which after formation were captured between the RER membranes, ×50 000

result, the detection of polysomes and interpretation of their relationship to the glycogen particles becomes difficult, if not impossible. The number of polysomes present in the glandular epithelial cells does, however, diminish after day 16 of the cycle. Differentiation between RNA and glycogen is possible with the use of selective staining of glycogen (Thiéry 1967) and differential staining of RNA (Bernhard 1968). During the initial stages of glycogen formation, polysomes which carry both glycogen and RNA particles are frequently present. The glandular epithelial cells do not, however, all respond simultaneously to the hormonal stimulus. During the immediate postovulatory period only a small number of cells begin to synthesize glycogen right away; the majority of cells start synthesis later, in ever-increasing numbers. During the early stages of glycogen synthesis another morphological proof of the relationship between RNA and glycogen deposition is occasionally found, namely RER coils. These consist of either spiral-like or circular cisternae of RER which are frequently continuous with the perinuclear space (Fig. 24). Between the coiled RER membranes narrow rims of cytoplasm are encircled. The outermost membranes, i.e., those facing the free cytoplasm are studded with ribosomes, while the membranes facing the encircled areas of cytoplasm are studded with glycogen particles. This glycogen consists mostly of α-particles.

3.1.3.2 Glycogen Discharge

Glycogen cannot generally permeate cellular membranes, it must be disintegrated into glucose molecules before it can be discharged from the cells. During the second third of the secretory phase of the cycle, particulate glycogen is frequently seen in the lumen of the endometrial glands (Figs. 25, 26). Because of the large amounts of glycogen synthesized and discharged from the glandular epithelial cells, the mechanism by which it is discharged is of particular interest. Some of the intraluminal glycogen is membrane bounded, some is not. Serial sectioning shows the membrane-bounded glycogen accumulations not to be intracellular glycogen, but truly separate apical areas of glycogen-filled cytoplasm. The actual process of constriction and release of apical cytoplasmic areas cannot be documented with static pictures of the form produced from ultra-thin sections. There are, however, numerous intermediate steps in this process of discharge which can be seen on almost every section; this facilitates visualization of "glycogen secretion" (Fig. 25).

The whole process of discharge corresponds to apocrine secretion. The first step is represented by an accumulation of apical glycogen which causes the apical cell membrane to bulge into the glandular lumen. The bulging areas of cytoplasm do not contain any organelles, except for a few vesicles. The glycogen present in the apical areas consists almost entirely of β-particles. During the phase of apocrine secretion of glycogen, the apical cell membranes appear to be very fragile. Frequently they have discontinuities, through which other constituents of the cells can pass into the glandular lumen. Thus, it is common to find even giant mitochondria inside the glandular lumen. The discharged membrane-bounded glycogen areas seem to dissolve their membranes, so that at the end of the second third of the secretory phase most of the glycogen

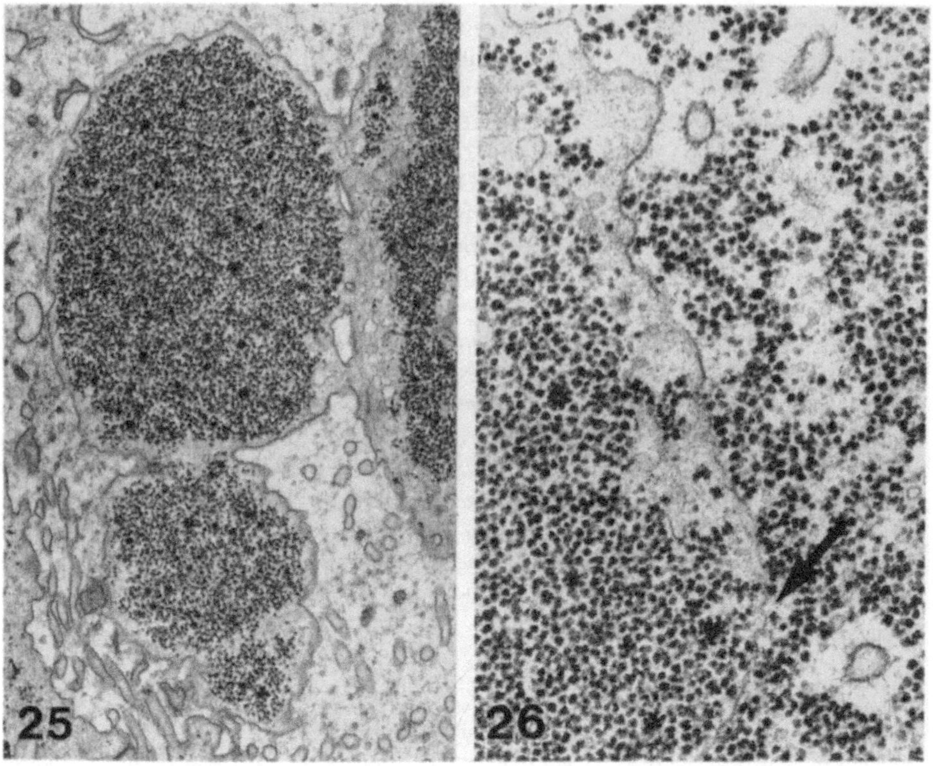

Figs. 25, 26. Glycogen discharge from glandular epithelial cells. **Fig. 25.** The budding-off of an apical glycogen-filled area of cytoplasm, ×15 000. **Fig. 26.** Intracytoplasmic and intraluminal glycogen packed so tightly that the cell membrane is difficult to depict. On the cellular side the glycogen is immediately adjoining the cell membrane as if it was going to be discharged in particulate form (*arrow*). The microvillar antennulae can be clearly distinguished due to distance from the β-particles, ×37 500

present in the glandular lumen is no longer bounded by membranes. When most of the glycogen has been discharged from the cells, the intraluminal glycogen sometimes becomes very tightly packed. Since the glycogen also keeps a distance from the microvilli equivalent to the length of the antennulae (Fig. 26), the microvillar antennulae are now more clearly visible.

3.2 Physiological Postovulatory Endometrium: Stromal Elements

3.2.1 Endometrial Granulocytes (K Cells)

3.2.1.1 General Characteristics

The morphology of endometrial granulocytes (or K cells) varies considerably during the course of a cycle. Therefore, whenever describing the morphology of the endometrial granulocyte, the exact date of the cycle from which the granulocyte was obtained should be specified. During the late secretory phase, i.e., between days 24 and 28, the endometrial granulocytes have attained a fully differentiated state. This is also the period when endometrial granulocytes are most numerous. In most cases, the number of endometrial granulocytes equals the number of predecidual cells. This can be clearly observed on scanning microscopic photographs, whereas transmission microscopic photographs do not usually reveal the distribution patterns of different cells, due to the unfavorable thickness of the sections. Fully differentiated endometrial granulocytes have an oval shape with a longitudinal diameter of 11 µm and a transversal diameter of 7.5 µm (Fig. 27). The cell surface is modulated by cellular extensions which vary in diameter between 0.2 and 0.5 µm, and reach a length of up to 6 µm. On sections these cellular extensions are hardly ever cut parallel to their long axis, so that they frequently look like normal microvilli. The most outstanding characteristic of the K cells are their secretory granules. On sections through the center of the cells an average number of 10–12 secretory granules can be found (Fig. 27).

3.2.1.2 Subcellular Organization

3.2.1.2.1 The Nucleus

Owing to the eccentric location of the nucleus, K cells have a pronounced polarity (Fig. 27). The nucleus is more or less kidney shaped, with the convex side facing the neighboring cell membrane. An almost organelle-free peripheral rim of cytoplasm is present in most cells and can be regarded as a characteristic feature. This rim is particularly pronounced when the cells are laden with glycogen. Even on superficial examination using TEM sections K cells can easily be detected because of their conspicuous nuclei. The nuclei of the endometrial granulocytes possess a broad rim of heterochromatin which is three to six times as wide as that found in other stromal cells (Figs. 27, 31). This heterochromatin rim is present irrespective of the type of fixation used; it can even be seen with simple OsO_4 fixation. Unlike stromal fibrocytes, K cells do not possess a nucleolus. Because of the convex-concave shape of the nuclei, they occasionally appear twice on the same section. In addition to those cells which have only one nucleus, there are also K cells which possess two nuclei, as can be seen from serial sections. Diplosomes are a constant feature of K cells. In addition to centrioles, GA-fixed tissue also revealed microtubules which extended radially from the diplosomes. Stages of mitosis can be found in all stages of differentiation of K cells. With the undifferentiated K cells from the proliferative and early secretory phase, mitotic stages can hardly be distinguished from those of stromal fibrocytes. Identification becomes easy, however, as soon as the K cells start to develop their typical secretory granules which are not degraded during mitosis (Fig. 28).

3.2.1.2.2 Glycogen

During days 24–28 of the cycle, when K cells are fully differentiated, most of them possess large amounts of glycogen. The glycogen is always present as β-

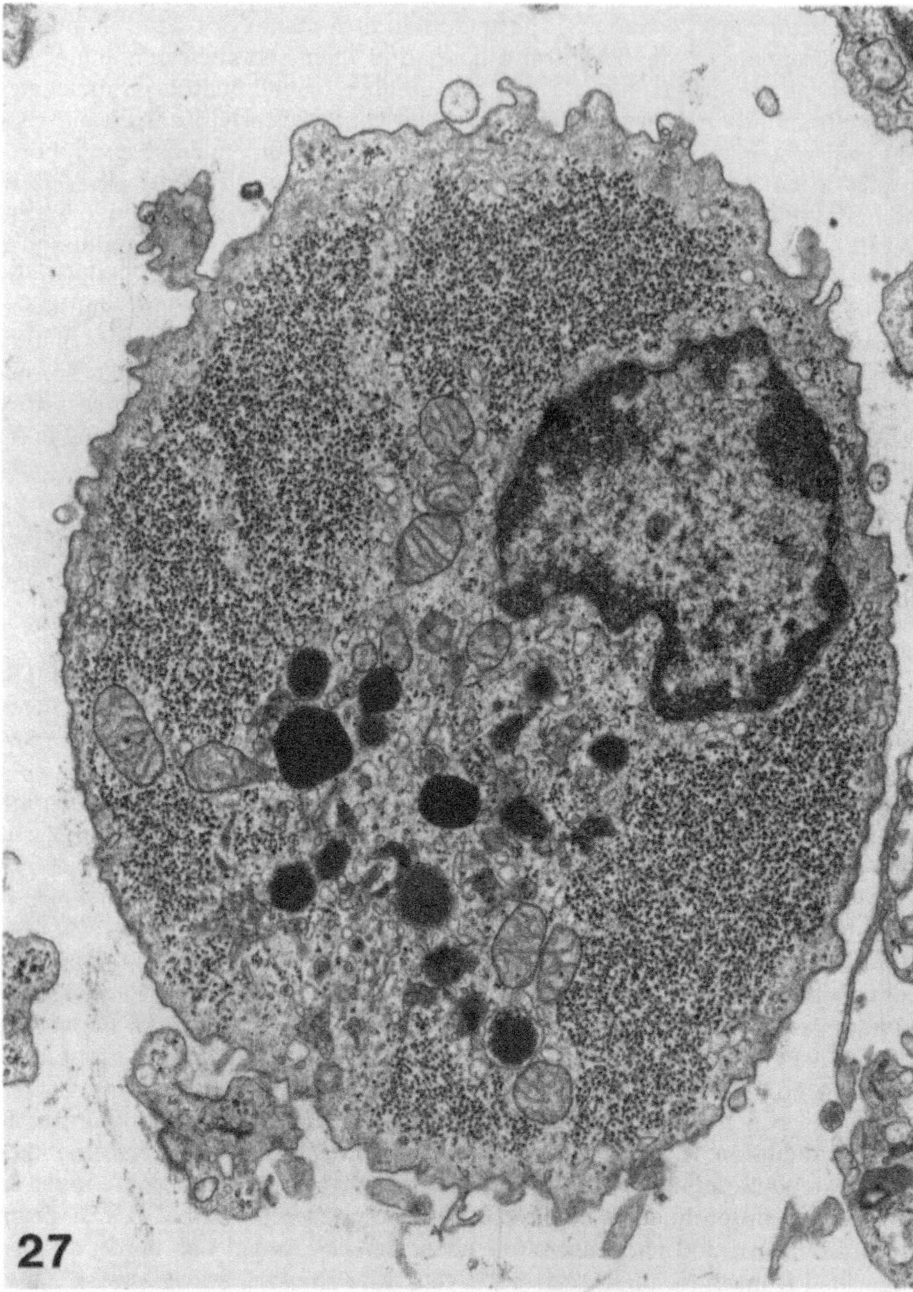

Fig. 27. Typical mature K cell with characteristic features of eccentric nucleus with a broad rim of heterochromatin, large amount of glycogen, and secretory granules, ×15000

particles. This is so very characteristic that it can be taken as a means of identifying K cells. Even small areas of tangentially sectioned cells can be identified using this criterion. Whenever α-particles are present, the cell in question cannot be a K cell. The majority of fully differentiated endometrial

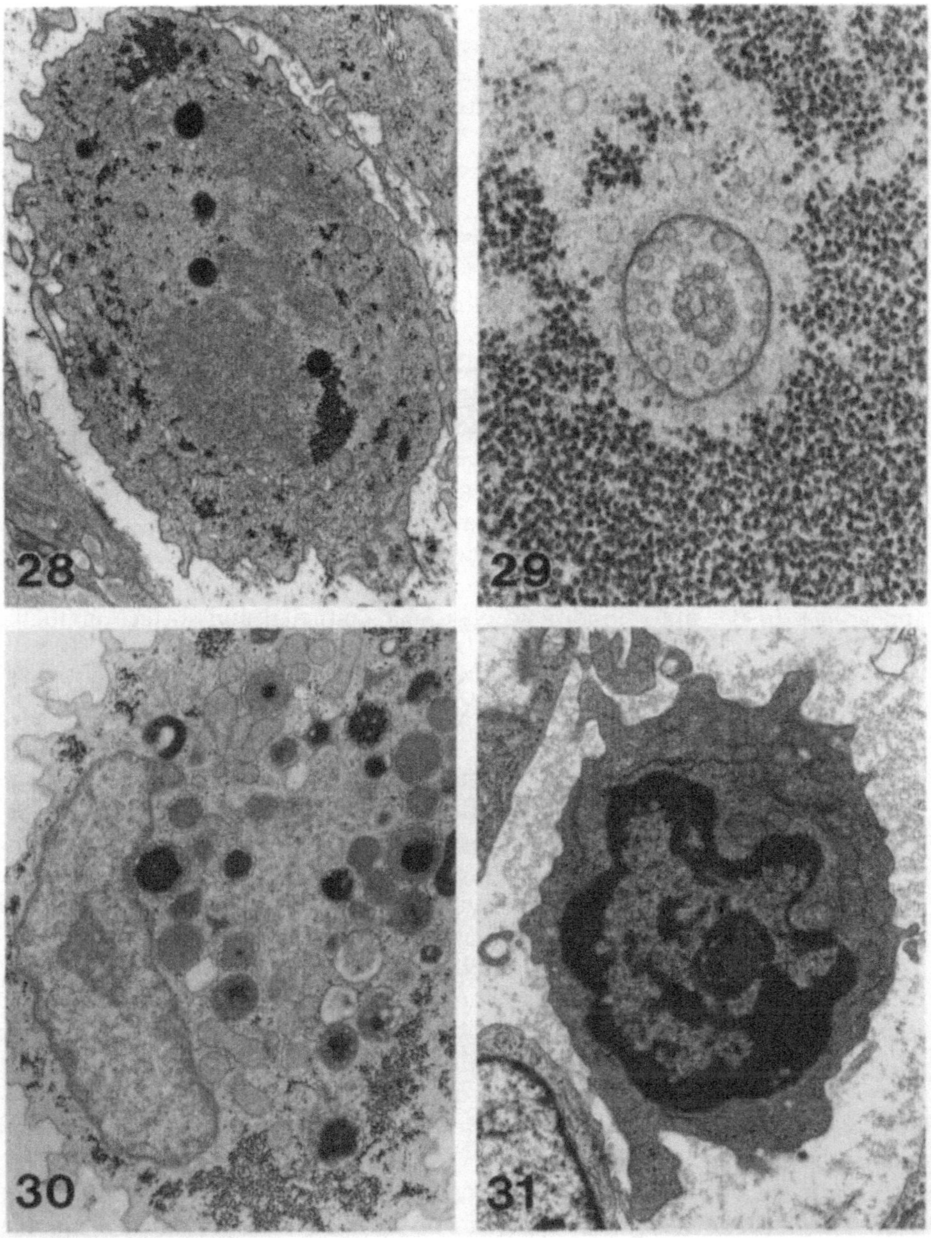

Figs. 28–31. K cells. **Fig. 28.** Mitotic K cell. Mitosis of K cells is frequently encountered during the early to middle secretory phase. Despite the partial dedifferentiation of these mitotic cells they still can easily be identified on the basis of their secretory granules, ×8200. **Fig. 29.** Multivesicular body, the precursor of K cell secretory granules, ×45 000. **Fig. 30.** Various stages of secretory granule formation are present in the same cell, ×8200. **Fig. 31.** K cell precursor from the 10th day of the cycle. These cells stand out from other stromal cells because of the broad rim of heterochromatin and the beginning formation of MVBs, ×9500

granulocytes possess a moderate amount of glycogen. Occasionally there are, however, cells present which are either absolutely devoid of glycogen or are completely filled with it. Both extremes are predominantly found during the first term of pregnancy or with progesterone-releasing IUDs.

3.2.1.2.3 ER, Golgi, Mitochondria

The organelles of K cells do not appear to be different from those of other cells. Apart from a moderate amount of RER, they possess numerous vesicles which either belong to the SER or are part of the Golgi apparatus. The Golgi apparatus consists of several seemingly highly active dictyosomes which are located in the cytoplasm facing the concave side of the nucleus. This is also the region where most of the above-mentioned vesicles are located. They range in diameter between 50 and 100 nm and the majority are electron lucent. A second category of vesicles is present in the area of SER, which are at least twice as large as the vesicles in the Golgi area.

The mitochondria of K cells are rather inconspicuous with respect to both their number and the structure of their cristae. They measure from 0.4 to 0.6 µm in diameter and are up to 1.5 µm long, making them very similar in appearance to the mitochondria of glandular epithelial and surface epithelial cells.

3.2.1.2.4 Secretory Granules

The most prominent feature of fully matured K cells is the presence of secretory granules. On a cross section through a mature K cell, about 10–12 secretory granules can be found (Figs. 27, 30), measuring between 0.4 and 1.5 µm in diameter. Occasionally they can even reach a diameter of up to 2.9 µm. The secretory granules are surrounded by a single membrane. After single fixation with OsO_4 they have a homogeneous electron dense content. Two different kinds of secretory granules can be found. In the first the limiting membrane is in direct contact with the content of the vesicles; in the second there is an electron transparent rim between the limiting membrane and the content of the vesicles which frequently contains vesicles of the size found in the Golgi area. The presence of this electron transparent margin is independent of the type of fixation used. Both types of vesicles can be found side by side in the same cell.

Precursors of the secretory granules are found predominantly in the Golgi area. The smallest recognizable precursors measure 150–300 nm in diameter. Larger vesicles usually contain numerous smaller vesicles (Fig. 29). During differentiation of the endometrial granulocytes, all intermediate stages of development of the secretory granules can be observed, ranging from vesicles in the Golgi area to secretory granules that, except for the outermost region, are filled with an electron dense content (Fig. 30). Myelin-like structures in the secretory granules are frequently encountered, even in fully matured K cells. Glycogen particles inside the secretory granules develop especially towards the end of the secretory phase of the cycle, shortly before desquamation. The glycogen is made up solely of β-particles.

During normal cycles, i.e., without implantation of a developing blastocyst, there is apparently no discharge of secretory granules into the intercellular space – secretory granules were never found in the intercellular space. Towards the end of the cycle the content of the secretory granules becomes less electron dense, and acquires a heterogeneous, almost foamy, appearance.

3.2.2 Origin of K Cells

The K cell described above is a fully mature endometrial granulocyte which can be found during the late secretory phase of the female cycle. Before this stage of maturity is reached, the stem cells pass through a series of developmental stages. Immediately following desquamation, a precursor cell can be found in the proliferating stroma. It is, however, not before ovulation that this type of cell occurs in greater numbers, and it cannot readily be identified as a stem cell of the endometrial granulocytes as it does not possess any granules. Only the meticulous comparison of different developmental stages in the formation of K cells finally enables the positive identification of the precursor cell during the early proliferative phase of the cycle. The first detectable precursors of endometrial granulocytes are oval cells which have a longitudinal diameter of 7–8 μm and a transversal diameter of 4–5 μm. Regular stromal fibroblasts are much larger, with a longitudinal diameter of 10–15 μm. Even at this early stage, the first detectable endometrial granulocytes have a contour very similar to that of mature K cells, and also have the same microvillous or pseudopod-like extensions. They possess very few organelles; on an average section through the central region of a precursor cell usually one minute dictyosome, a single RER cisterna, and about 4–6 mitochondria are detected. Neither secretory granules, nor their multivesicular precursors are present. The cytoplasm stains very strongly due to the large number of ribosomes, which are mostly single ribosomes (Fig. 31). The nucleus of these early endometrial granulocytes shows a most striking difference to that of normal stromal fibroblasts; it has a broad electron dense rim of darkly staining heterochromatin which is independent of the type of fixation used. The typical kidney-shaped form of the nucleus is, however, not present in early precursor cells.

3.2.3 K Cells Under a Progesterone IUD

Because of the scarcity of very early human implantation stages available for investigation, it is absolutely impossible to carry out a planned study on the morphology of K cells during early pregnancy. For this reason the decidualization of the endometrium caused by a progesterone-releasing IUD is the ideal model for the study of early changes in K cells. During the first postinsertion months K cells are frequently in direct contact with the differentiating predecidual cells. K cell extensions can be seen within invaginations of the predecidual cell membrane (Fig. 32), frequently connected to the invaginated membranes by means of contacts similar to gap junctions. Further differentiation does not occur in both cell types before these contacts have taken place.

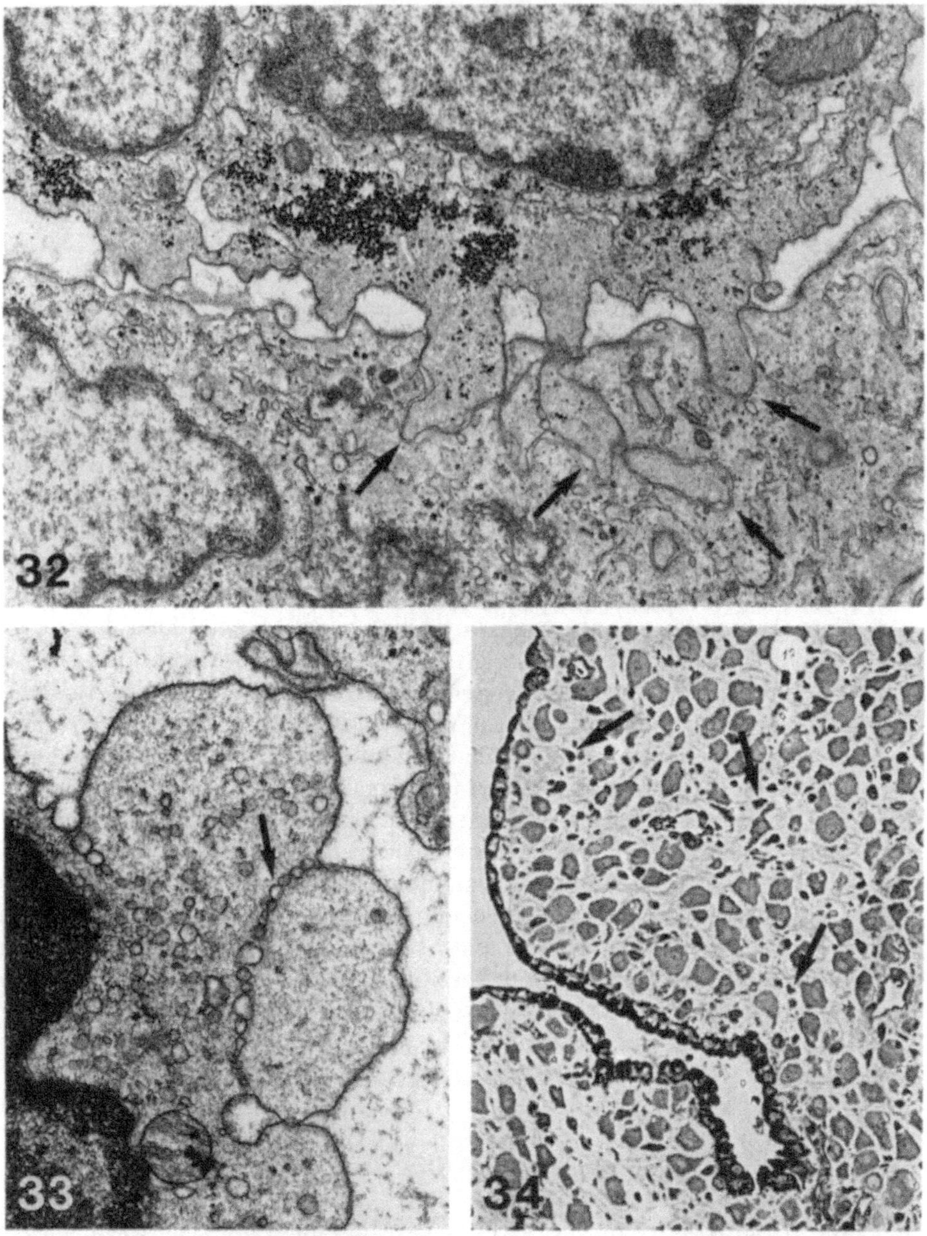

Figs. 32–34. K cells. **Fig. 32.** K cell in contact with a predecidual cell (*arrows*). Between K cell extensions and the wall of predecidual cell invaginations gap junctions are temporarily formed, ×16 000. **Fig. 33.** Vesicular demarcation line formed at the base of a bud of cytoplasm which apparently is being detached from the K cell (*arrow*), ×12 000. **Fig. 34.** Semi-thin section of a decidualized area from a patient with a progesterone IUD. The K cells (*arrows*) have discharged most of their secretory granules, ×220

With more differentiated cells, contacts between endometrial granulocytes and predecidual or later decidual cells can no longer be observed. After usage of the progesterone-releasing IUDs for more than 3 months the K cells start to degranulate, i.e., they unload their secretory granules into the intercellular space. Concomitant with the discharge of the secretory granules, part of the cytoplasm of the K cells is discharged through constriction of peripheral areas of cytoplasm (Fig. 33). This constriction cytoplasm is preceded by the protrusion of a cytoplasmic extension. At the base of the protrusion, a line of vesicles accumulates, forming a demarcation of the area to be discharged (Fig. 33). In slightly further developed cells, the vesicles have merged and formed a continuous line. Concurrently with the formation of vesicular demarcations and subsequent budding off, isolated cytoplasmic areas (as verified by serial sectioning) appear in the vicinity of K cells. The isolated cytoplasmic areas are frequently filled with glycogen of the same type that is present in the mature endometrial granulocytes. After prolonged use of progesterone-releasing IUDs, i.e., more than 3 months, the number of cells similar to the precursor cells rises again (Fig. 34). After 2–3 months of usage of the IUDs, the tissue areas in direct contact with the IUD are practically devoid of K cells containing secretory granules. Tissue areas which are not in direct contact with the IUD show a progressive degranulation, depending upon the distance from the IUD and the length of exposure to it. But even after 2 or 3 years of usage (with two or three successive IUDs) the regions of the uterotubal junction and the fundus still possess a certain number of endometrial granulocytes.

3.2.4 K Cells During Pregnancy

As stated above, K cells from very early implantation stages are almost impossible to study. K cells from 6 to 8 weeks of gestation are, however, more frequently available for investigation, due to planned abortions. The earliest implantation stage studied was that of a 3 mm crown-rump (CR) length embryo. Even at this early stage, it is evident that K cells degranulate and give off peripheral areas of cytoplasm filled with glycogen. As pregnancy progresses K cells of the basal decidua degranulate almost totally and reduce their size by the detachment of peripheral areas of cytoplasm. In tissue of decidua basalis from the end of the 3rd month of pregnancy, almost all of the K cells have discharged their secretory granules and have reduced their size. Thus, it becomes almost impossible to detect K cells on light microscopic sections. On electron microscopic sections K cells can occasionally still be found during the 5th month of pregnancy; identification of K cells is still possible at this stage on the basis of the few secretory granules still present. Another characteristic feature of "aged" K cells is the morphology of the nucleus which is identical to that of mature K cells (Fig. 35).

3.2.5 Ultimate Fate of K Cells

During the normal menstrual cycle which does not result in pregnancy, K cells hardly ever discharge their secretory granules, nor do they reduce their

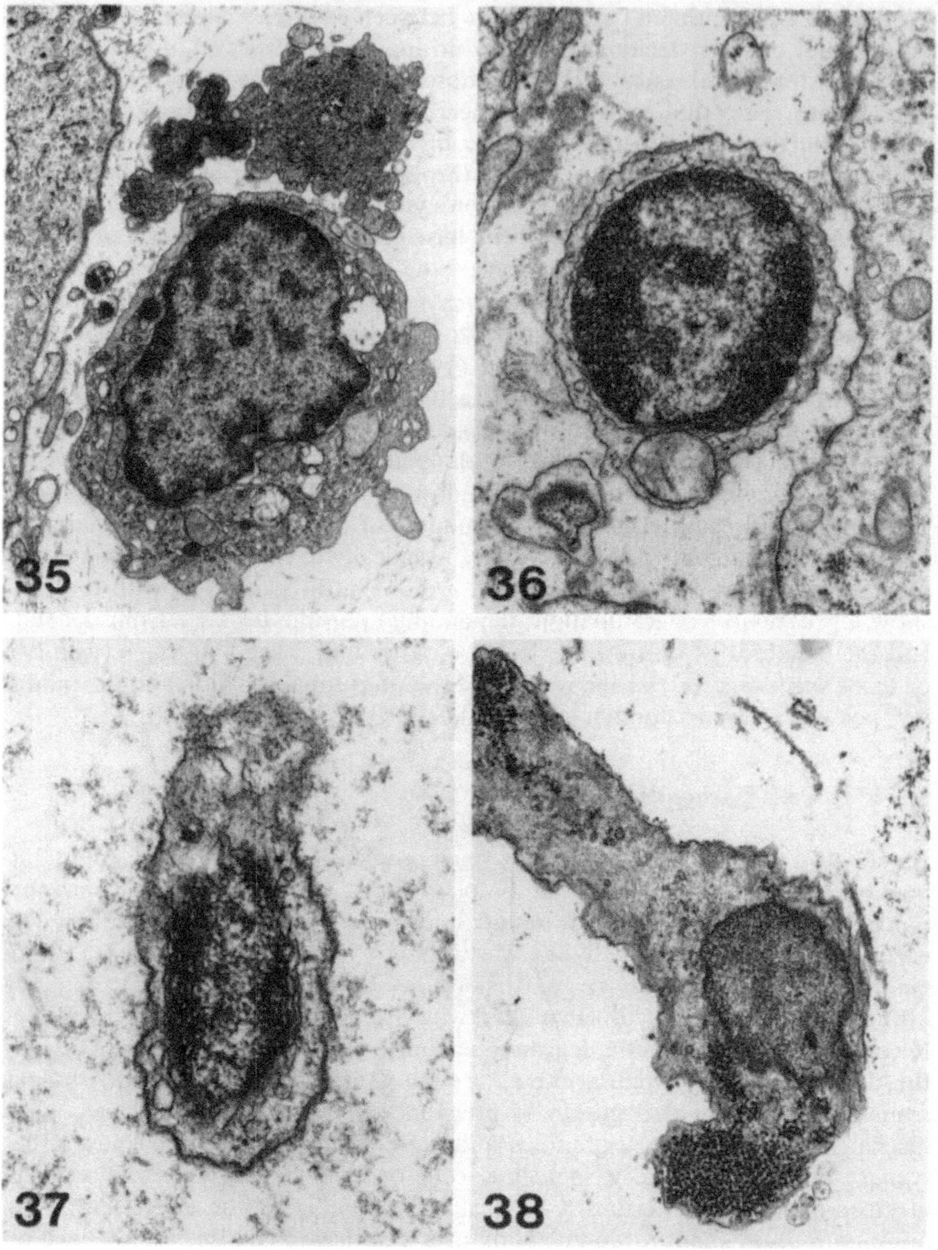

Figs. 35–38. Fractionating K cells. **Fig. 35.** K cell from the 2nd month of pregnancy. Except for two, the secretory granules have been discharged from the cells. Glycogen-filled parts have been detached from the cells (*top*), ×16500. **Fig. 36.** Through detaching of cytoplasm and fractionation of the nucleus miniature cells are formed. A solitary mitochondrion has remained in the cell, ×21000. **Fig. 37.** Extremely small miniature cell, magnified almost 70% more than cell in **Fig. 36.** ×35000. **Fig. 38.** Fractionated piece of nucleus inside a cellular extension just prior to being detached, ×14200

size through detachment of peripheral areas of cytoplasm. K cells are mostly located within the stratum functionale, i.e., the upper part of the endometrium which is lost during menstruation. Thus, the majority of K cells are desquamated and lost with the menstrual blood.

With the pregnant uterus the fate of K cells is totally different. Nevertheless, K cells which are already difficult to detect at the end of the 4th month of pregnancy have almost totally disappeared at the end of the 5th month. The cells appear to continue to detach parts of their cytoplasm (Fig. 38), whereby they are transformed into miniature cells (Fig. 36). A closer look reveals that with progression of pregnancy the endometrial granulocytes not only detach parts of their cytoplasm but also parts of their nuclei which apparently fragment during involution of the cells. Due to the final size of the miniature K cells they cannot be detected on normal histologic section. They measure about 1.6–2.4 μm in diameter, with a nucleus that ranges between 1.1 and 1.9 μm (Figs. 37, 39). The miniature nuclei of these cells are surrounded by a narrow rim of cytoplasm which contains very few organelles. Occasionally mitochondria of regular size are present and cause the plasma membrane of the cells to bulge (Fig. 37). At the end of pregnancy even the miniature cells have vanished from the decidua.

3.3 The Endometrium Under a Progesterone-Releasing IUD

3.3.1 General Aspects

There are two target tissues in the uterus on which the progesterone released from the IUD exerts its action, namely the glandular epithelial cells and the stroma. The effect on both tissues differs greatly. Furthermore, the effect of the hormone on the endometrium depends upon the localization of the tissue with respect to the IUD. The following description of the morphology of the endometrium under the progesterone IUD covers both aspects.

3.3.2 Usage up to 3 Months: Stroma in Direct Contact with the IUD

Light microscopic preparations show that the tissue which is in direct contact with the vertical reservoir of the progesterone IUD is already decidualized by the end of the first postinsertion month. This decidualization is caused by the released progesterone and does not involve the formation of periarterial coats as is typical in physiological decidualization. The wave of decidualization starts immediately in the tissue which is in direct contact with the IUD and is, therefore, initially restricted to the superficial stroma; it does not even cover the entire compacta. This initial decidualization occurs independent of the date of IUD insertion, i.e., it is also present in cases where the IUD was not inserted at the end of menstruation but, for instance, during the middle of the cycle. To study the actual process of decidualization during different stages, tissue from only one patient may be used since the various stages of decidual differentiation will most certainly be demonstrated. The degree of decidualiza-

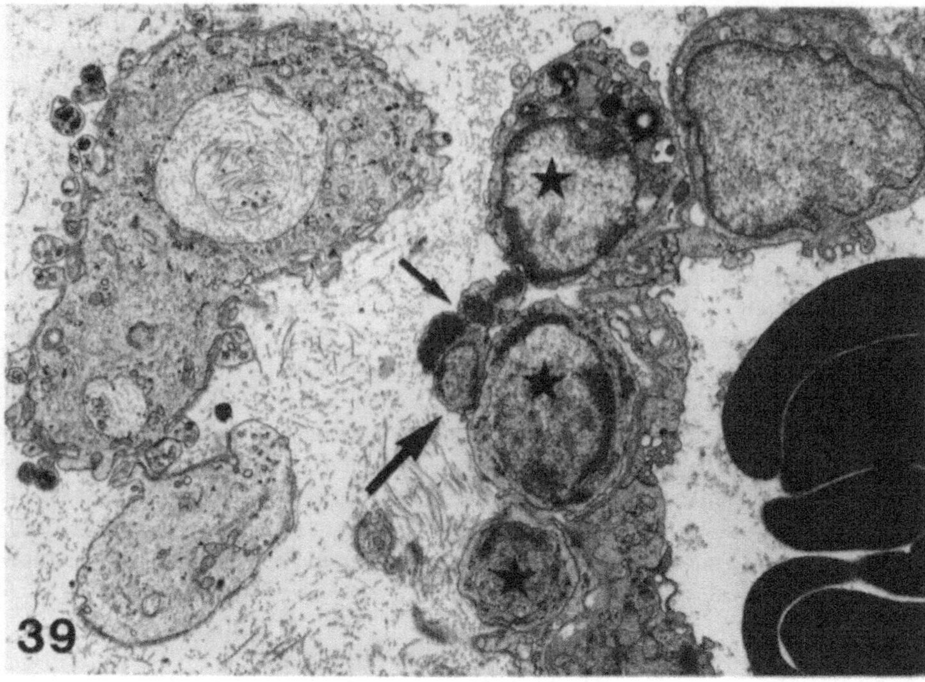

Fig. 39. K cells from the 3rd month of pregnancy (*asterisks*). The cells are in direct contact with the wall of a venule. A miniature cell has been formed already (*large arrow*) which appears in the process of detaching a glycogen-filled area of cytoplasm (*small arrow*), ×6000

tion depends upon the distance of the tissue from the IUD; the closer the tissue to the IUD, the more pronounced the degree of decidualization. Thus, during the initial phases, different stages of decidualization can be found side by side.

On electron microscopic sections the first indications of the onset of decidualization are easily seen: the undulations in the nuclear membrane of the stromal fibrocytes disappear and the nuclei increase their volume. This is followed by growth of the whole cell (Fig. 40). Concomitantly, the synthetic apparatus of the cells proliferates – particularly the Golgi apparatus and the RER. Subsequently the free ribosomes and polysomes are augmented and the mitochondria multiply. The first newly formed mitochondria have normal dimensions (0.3–0.4 μm), and their cristae do not differ from that of other mitochondria. As a result of these changes in stromal cell morphology the fibrocytes have differentiated into predecidual cells. The endpoint of this development is marked by the occurrence of glycogen stores, which contain α- as well as β-particles, distributed in small accumulations over the whole cell. The predecidual cells formed by the action of the hormone released from the IUD cannot be distinguished from predecidual cells formed under physiological conditions at the end of a cycle. There is one fact, however, that must be mentioned at this point: under the progesterone IUD fully decidualized (not only predecidualized) cells are already present at the end of the first postinsertion month; these cells are formed in the tissue that is in direct contact with the

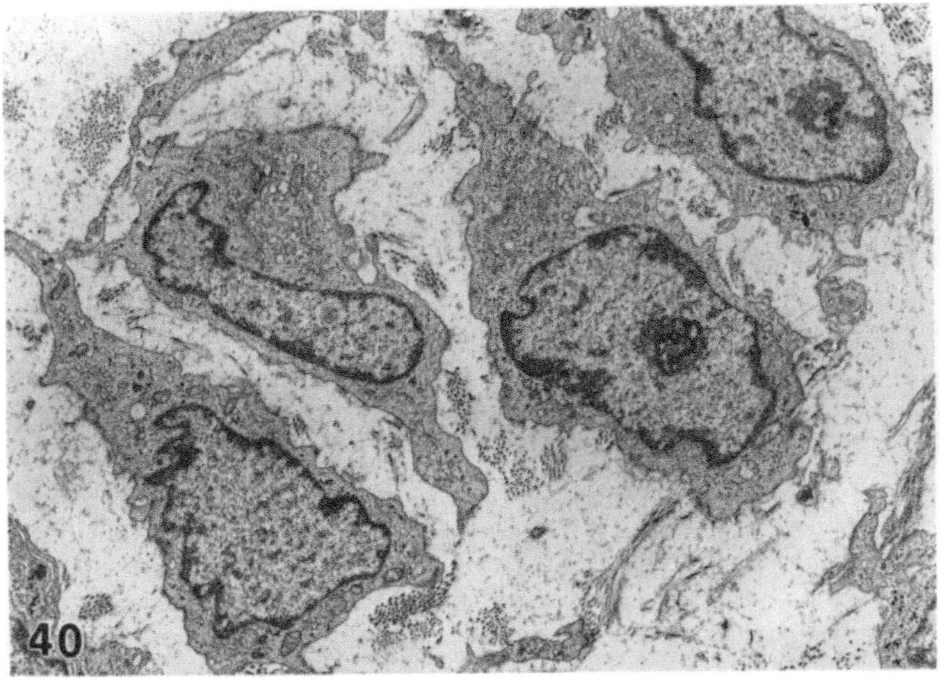

Fig. 40. Predecidual cells from the 2nd month under a progesterone IUD. When compared with the precursor cell (the stromal fibroblast) it becomes obvious that the synthetic apparatus of the cells has been activated, the nuclear surface has become smoother, and nuclei are enlarged, ×6000

IUD. In this tissue zone there are no predecidual cells left at the end of the first postinsertion month.

3.3.3 Usage up to 3 Months: Stroma not in Contact with the IUD

During the first 3 postinsertion months the stromal areas which are neither in direct contact with the IUD nor at least nearby, show hardly any detectable morphologic changes. Some of the cells in these areas lose their nuclear membrane undulations and start to increase their nuclear volume. In the tissue areas in the immediate vicinity of the IUD (but not in direct contact) predecidual cells are formed, not only at the end but also during the proliferative phase of the cycle. This reaction, which is not correlated to the cycle, is limited to the immediate vicinity of the IUD. In tissue areas distant to the IUD, the cells react in correlation with the uterine cycle; at this point of IUD usage they are definitely not influenced by the progesterone from the IUD.

3.3.4 Usage up to 12 Months: Stroma in Direct Contact with the IUD

Unlike predecidual cells, fully decidualized cells differentiate through the action of the progesterone released from the IUD. The first developmental

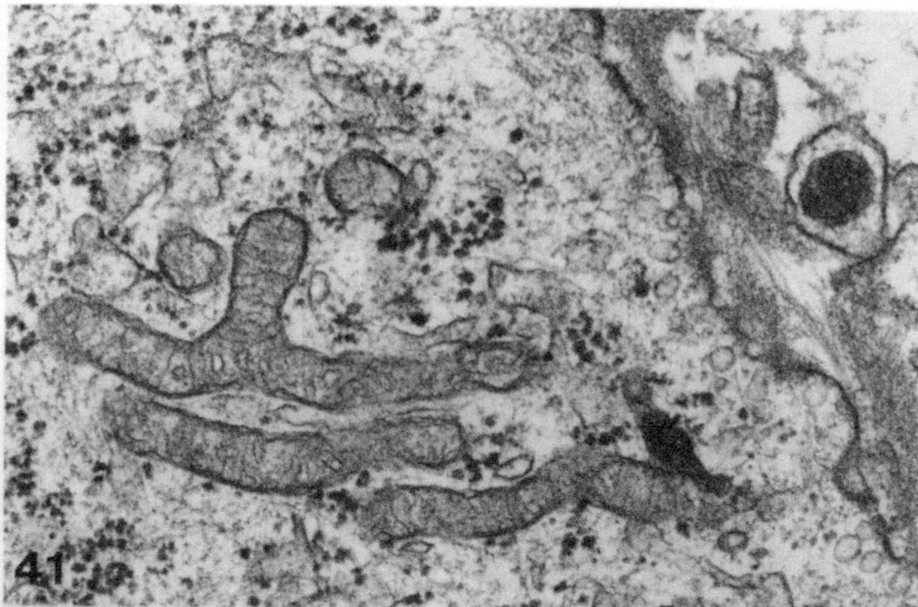

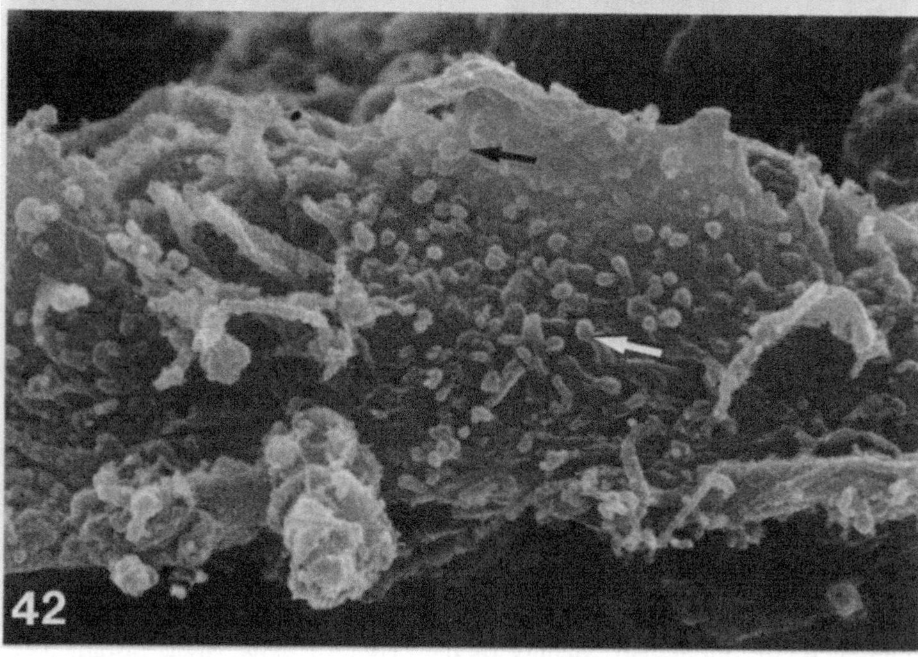

Figs. 41, 42. Decidual cells. **Fig. 41.** Through budding and fission the mitochondria of decidual cells multiply and reduce their size down to one-tenth of the volume of normal mitochondria. Note the BL coat and the decidual granule on the *right*, ×44 000. **Fig. 42.** Scanning electron micrograph of part of a decidual cell. Numerous decidual granules are visible (*arrows*), ×4800 step towards decidualization is the formation of large quantities of glycogen which are no longer disseminated, but form large aggregates of glycogen particles. The glycogen consists exclusively of α-particles which are so tightly packed that other organelles are excluded from the glycogen areas. Subsequent to glycogen formation, the Golgi apparatus proliferates. On an average section taken from this period, up to 15 individual dictyosomes are present; these dictyosomes are spatially related to the RER. With progressive decidualization the Golgi apparatus becomes more and more active, which is reflected by the high number of Golgi vesicles present in the vicinity of the dictyosomes. In slightly further developed cells, vesicles in the Golgi areas can be found which are apparently coalescing to form multivesicular bodies (MVBs). As the Golgi apparatus is activated, the RER expands over the whole cytoplasm. The ER cisternae are slightly extended at this point, and have lost their ribosomes in the regions neighboring the dictyosomes. During the later stages of decidualization the RER loses almost all its ribosomes and starts to break up into a vesicular form possessing only a few ribosomes. The mitochondria multiply by means of budding and fission (Fig. 41), and their number is dramatically increased – a fact which almost escapes attention since the decidual cells themselves increase their volume by a factor of approximately 40. The volume density of mitochondria is, therefore, unchanged. As the mitochondria multiply, the individual mitochondrial volume drops to about one-tenth of the original volume. This is due to the rather limited growth of the mitochondria following fission and budding. The volume of the individual mitochondrion from the stromal fibroblasts is approximately 1.1 μm^3 compared to a volume of 0.1 μm^3 for the mitochondria from fully decidualized cells (Fig. 43). The degree of miniaturization of the mitochondria can in fact be taken as an indicator of the degree of decidualization. Since on the one hand the mitochondrial volume density of fully decidualized cells is more or less equal to that of predecidual cells, and on the other hand, the volume of the decidual cells has increased by a factor of 40, the total mitochondrial volume per cell increases drastically during decidualization (see Sect. 3.6).

During formation of the miniature mitochondria, the glycogen deposits are degraded and the α-particles vanish totally from the cell, leaving only a few β-particles remaining in the cells. Subsequently the Golgi apparatus is highly activated. The number of Golgi vesicles with both electron lucent and electron dense granules increases and the MVBs become more numerous. On the basis of their content the Golgi vesicles can be divided into two groups, one with a homogeneous electron dense content, and the other with a rather electron transparent, vesicular content. The size of MVBs ranges from 0.8 to 2.0 µm. The vesicles inside the MVBs measure about 40 nm in diameter and they are embedded in a fine granular matrix. In slightly further developed decidual cells, the MVBs can be found in the peripheral cytoplasm where they appear to be secreted from the cells by means of apocrine secretion. However, not all MVBs are completely secreted; many of them remain attached to the cells by a narrow stem, or cytoplasmic protrusion. Owing to the apocrine mode of secretion, MVBs are surrounded by a narrow rim of cytoplasm which is sometimes so narrow that the MVBs appear to be surrounded by a triple membrane. The aforementioned protrusions containing MVBs are one of the most typical

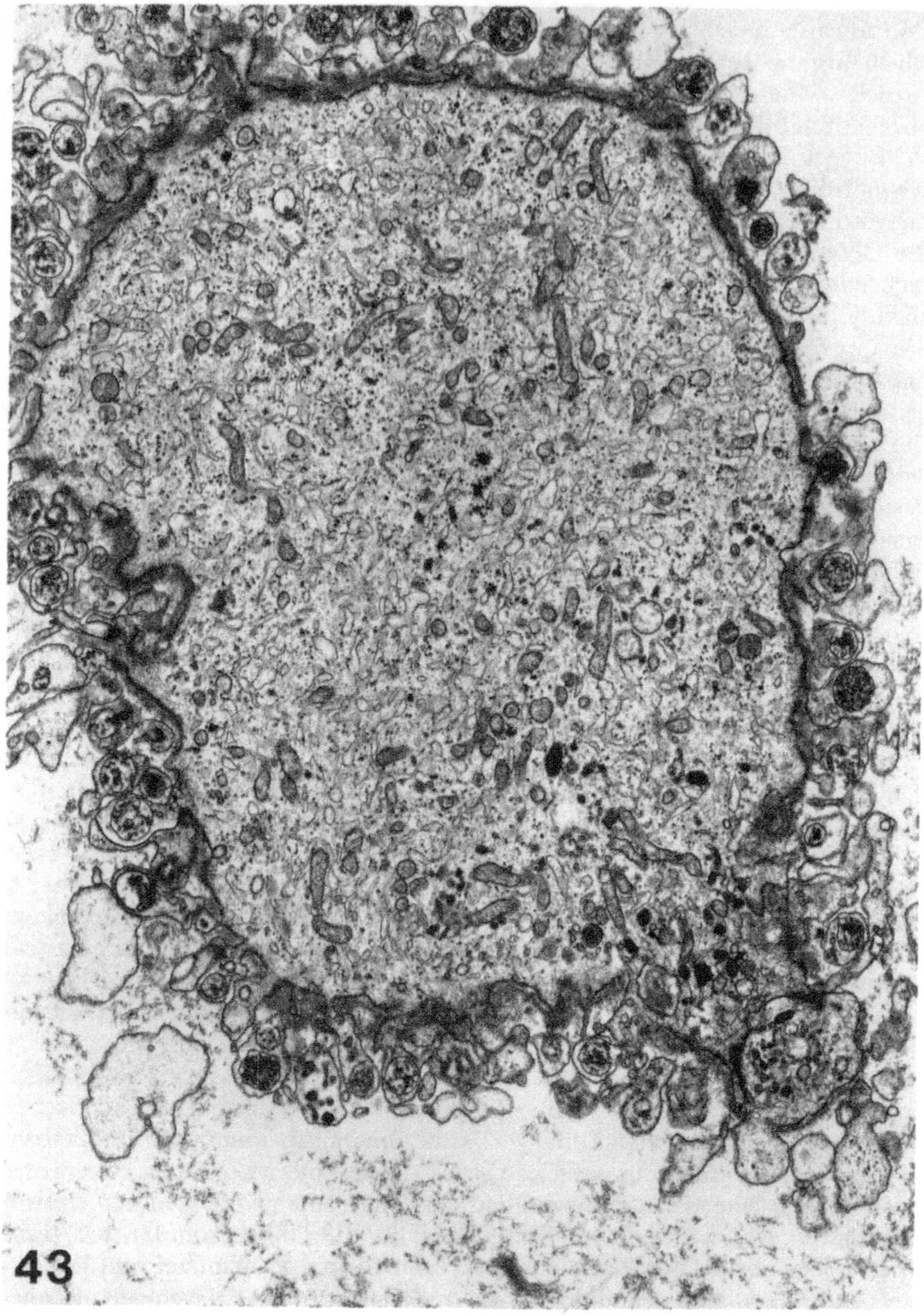

Fig. 43. Fully decidualized cell. Three features of decidual cells are clearly visible on this micrograph: decidual granules, miniature mitochondria, and the BL, ×13200

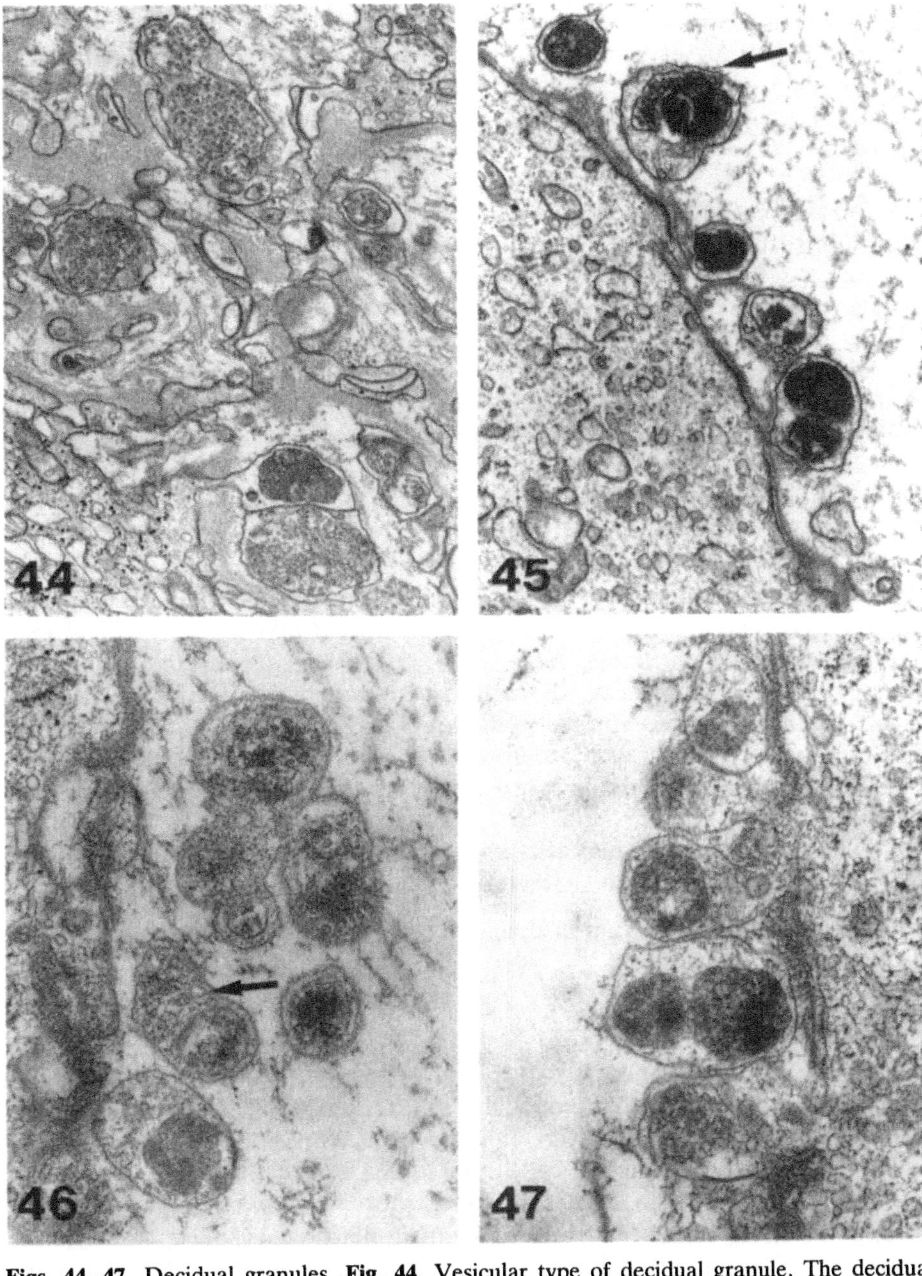

Figs. 44–47. Decidual granules. **Fig. 44.** Vesicular type of decidual granule. The decidual granule on top has a discontinuous limiting membrane which is suggestive of a discharge of the vesicles into the intercellular space, ×18000. **Fig. 45.** Osmiophilic type of decidual granule. One decidual granule contains both vesicles and osmiophilic matter (*arrow*), ×18000. **Fig. 46.** Decidual granules after digestion with pronase for 2.5 h, which removes part of the matrix, but leaves the vesicles (*arrow*) and osmiophilic substance intact, ×30000. **Fig. 47.** Decidual granules after digestion with α-amylase for 3 h, which leaves them basically unaltered, ×30000

features of decidual cells and, for this reason, have been named "decidual granules" (Spornitz et al. 1983). Mature decidual cells are covered with a coat of decidual granules which is particularly evident on SEM photographs (Figs. 42, 43). These decidual granules are of two types: (a) those with a more or less vesicular content (Fig. 44), and (b) those with a homogeneous, electron dense content (Fig. 45).

Enzymatic digestion performed with proteolytic enzymes (pronase) on ultra-thin sections only removes the matrix of the vesicular decidual granules. The vesicles themselves and the homogeneous content of the electron dense decidual granules is not digested (Fig. 46). Treatment with α-amylase digests the decidual granules only to a very limited extent (Fig. 47). Specific staining for carbohydrates in general, like staining for glycogen in particular, does not produce any contrast enhancement in the decidual granules (Fig. 48). Osmium tetroxide fixation, however, reveals that the decidual granules contain a relatively high amount of unsaturated fatty acids (Fig. 45).

Final maturation of stromal cells into decidual cells is achieved through the formation of a basal lamina (BL)-like coat. Initially this coat is discontinuous and covers only part of the cell surface, but fully mature decidual cells are covered by a continuous coat of BL-like material. The decidual granules are not covered, except for the stalk which connects them to the decidual cells. The BL ranges in thickness between 70 and 200 nm on most decidual cells. Occasionally it is as thick as 600 nm. In cases where the decidual cells are in direct contact with the surface epithelium or the glandular epithelium, the BL of both cell types is continuous. Not only the thickness but also the morphology of the BL is identical to that of epithelial and decidual cells. (see Sect. 3.9).

Fully developed decidual cells induced through the progesterone-releasing IUD thus have four unique characteristics:

1. The cell size (up to 40 µm in diameter)
2. Miniature mitochondria
3. A continuous coat of BL
4. Decidual granules

These features can already be found at the end of the first postinsertion month in those stromal areas which are in direct contact with the progesterone IUD. After 1 year of usage (recommended time for maximum usage, before change) practically all stromal cells in the areas of direct contact with the IUD and, therefore, likely to be decidualized, show the typical decidual features, and, immediately after formation, the first decidual granules; they can be observed to be discharged into the intercellular space. Serial sections provide unequivocal evidence of this discharge. With further progression of the decidualization of the endometrium, caused by continuing contact of the endometrium with the IUD, more and more decidual granules are found which after their discharge into the intercellular space, start to disintegrate and release their contents (Fig. 49). The size of these decidual granules ranges from 0.4 to 0.9 µm. The limiting membranes of the decidual granules are dissolved and the content of the decidual granules becomes part of the intercellular matrix (Fig. 50). Morphologically the intercellular matrix resembles the fibrinoid

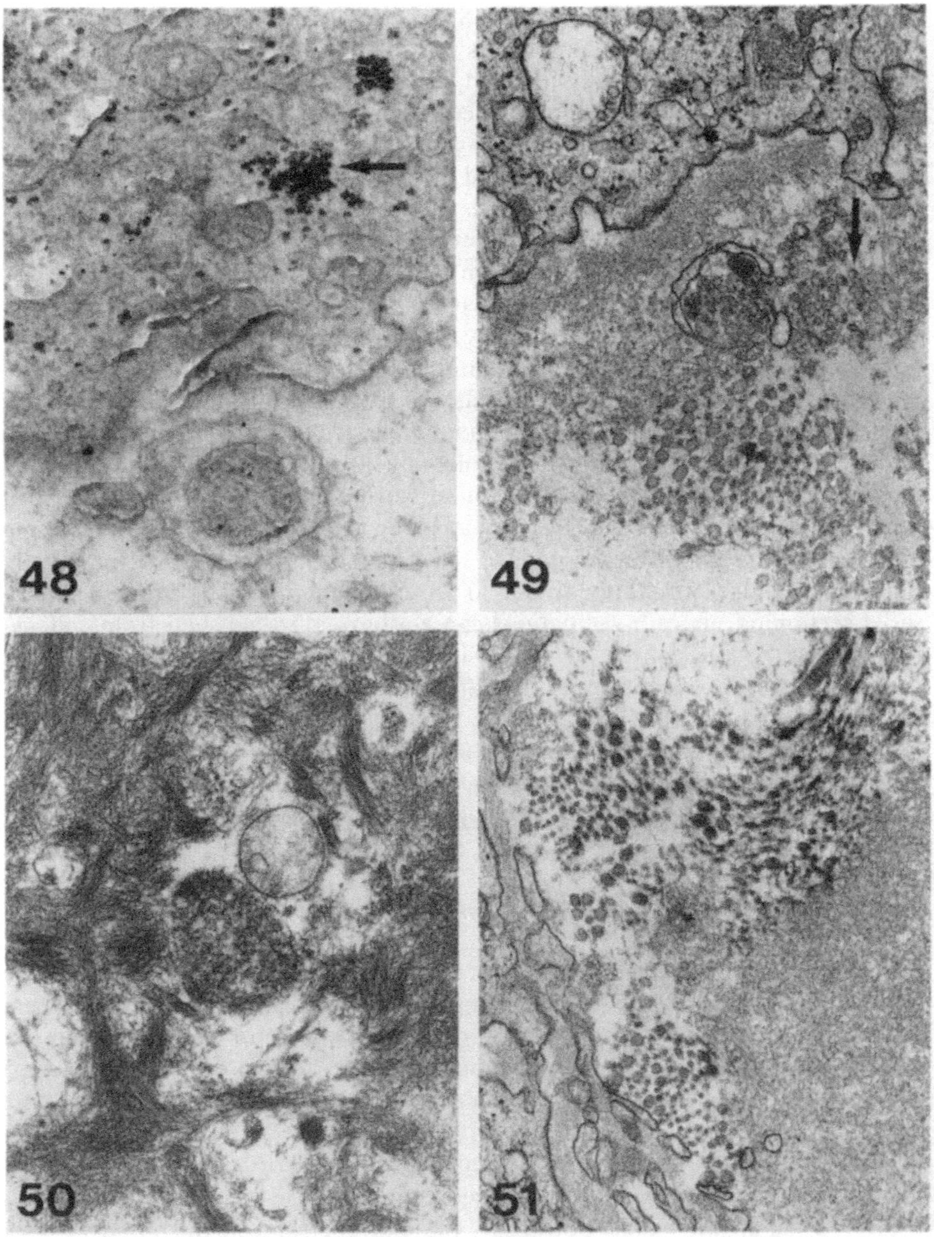

Figs. 48, 49. Decidual granules. **Figs. 50, 51.** Fibrinoid. **Fig. 48.** Decidual granule stained specifically for carbohydrates. The glycogen particles are well contrasted (*arrow*); the decidual granule (*bottom*) is left basically uncontrasted, ×30 000. **Fig. 49.** Decidual granule with apparent discharge of its content. Several vesicles are lying free within the intercellular space (*arrow*). In the *lower third*, there are cross-sectioned collagen fibrils, ×35 000. **Fig. 50.** Dissolving decidual granule which seems to be incorporated into the surrounding fibrinoid, ×26 000. **Fig. 51.** There appears to be a distinct demarcation line between collagen fibrils (*top and left*) and fibrinoid (*lower right*), ×26 000 of the uteroplacental interphase (see Sect. 3.5). In the areas where there is a massive discharge of decidual granules into the intercellular space, the amount of collagen fibrils is drastically reduced, as can be seen from TEM photographs and by means of silver impregnation according to Gomori. There appears to be an inverse relationship between collagen and fibrinoid. With progression of decidualization the amount of fibrinoid increases, while the amount of collagen drastically decreases. The borderline between collagen accumulations and fibrinoid areas is a very sharp one; they seem to exclude each other. Collagen fibrils are almost never embedded in a matrix of fibrinoid (Fig. 51). Fibrin is, however, frequently found to be associated with fibrinoid, particularly in areas where, due to vascular damage, extravascular red blood cells are present.

A characteristic feature of pronounced decidualization is the so-called collagen vacuole which is found in fully decidualized stromal areas. Collagen vacuoles are intracellular membrane-bounded vacuoles which contain collagen fibrils (Fig. 52). Many of these vacuoles are located in a juxtanuclear position. Similar vacuoles can also be found during the late secretory phase in predecidual cells. The collagen vacuoles of decidual cells are not simple invaginations as might be presumed; serial sections always prove them to be fully intracellular structures. Even in cases where the cells are not cut by serial sectioning, mere invaginations can easily be distinguished from collagen vacuoles on the basis of the BL lining. Unlike collagen vacuoles, undulations of the cell membranes and deeper invaginations are always lined by a BL (Figs. 52, 53). After prolonged usage of the IUD the collagen vacuoles are increasingly filled with lysosomal elements. Subsequently the collagen fibrils are disintegrated. Occasionally a discharge of the content of digested collagen vacuoles into the intercellular space can be observed.

3.3.5 Usage up to 12 Months: Stroma not in Contact with the IUD

Stromal areas which are not in direct contact with the progesterone-releasing stem of the IUD show the same morphological changes, but over a longer time course. This time lag is directly related to the distance of the tissue from the IUD. Stromal areas within the immediate neighborhood of the vertical stem of the IUD are decidualized first; stromal areas within the vicinity of the tubal openings are among the last to be decidualized – sometimes they are not decidualized at all. The amount of decidualization varies greatly from patient to patient and is apparently independent of the time of usage of the IUD. When decidualization does not occur, the endometrium of this area, i.e., the area of the tubal openings, responds normally to the varying levels of circulating endogenous hormones.

3.3.6 Usage for over 1 Year

The reservoir of progesterone in the vertical stem of the IUD contains enough progesterone to last for more than 1 year of usage. Thus, the effectiveness of the IUD is not impaired even after 15 or 16 months of usage. Although

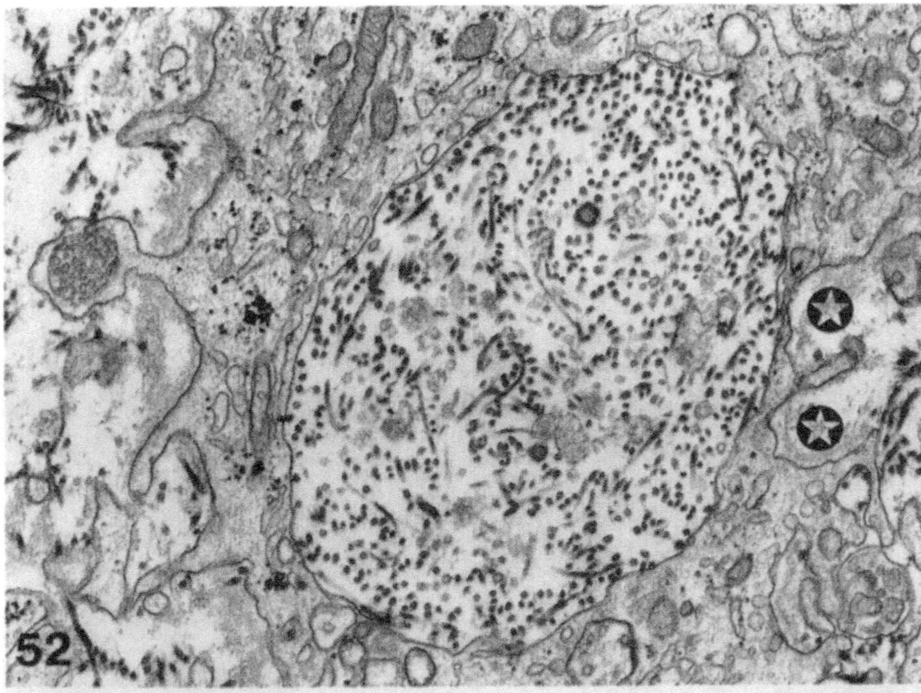

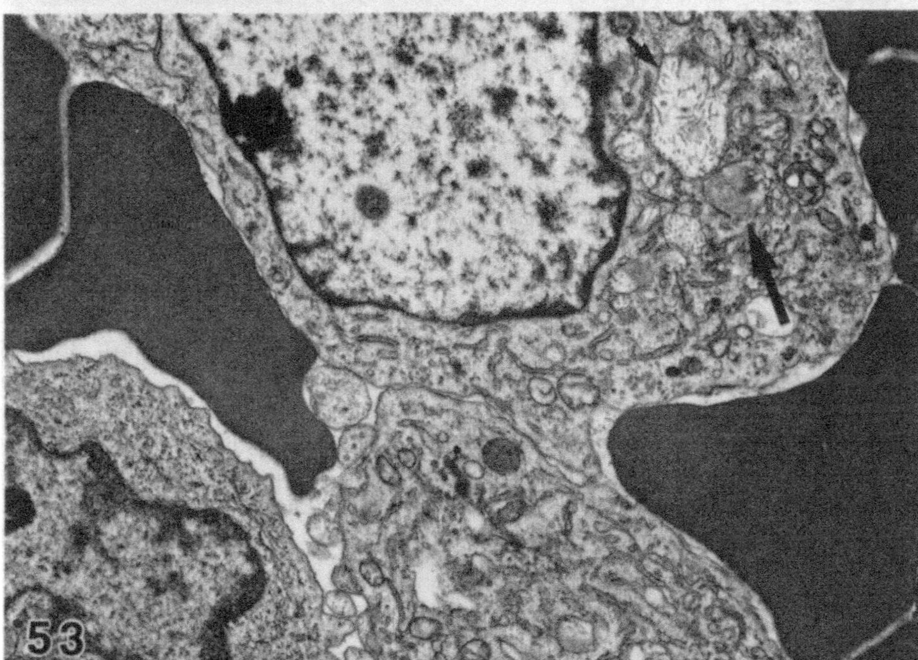

Figs. 52, 53. Collagen vacuoles. **Fig. 52.** Typical collagen vacuole of a mature decidual cell. Unlike invaginations (*asterisks*) collagen vacuoles are not lined by basal lamina, ×25000. **Fig. 53.** Decidual cells and a K cell within an area of extravasal blood. Because of the blood the cells are totally cut off from the stroma. A collagen vacuole (*small arrow*) and an invagination (*large arrow*) are present. The invagination is lined with BL; the collagen vacuole is not, ×9500

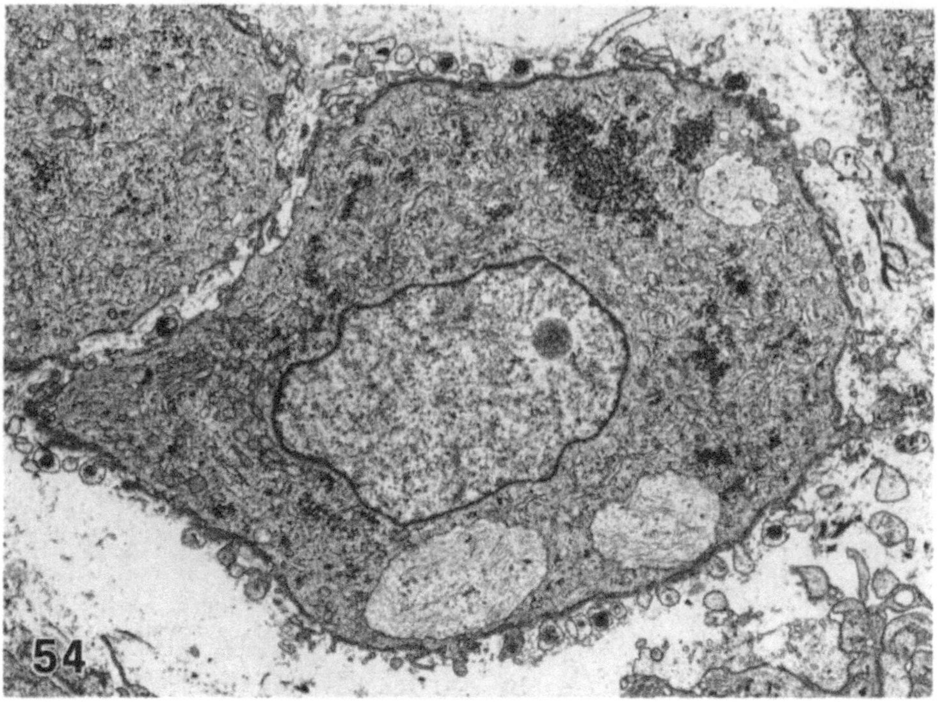

Fig. 54. Decidual cell induced through a progesterone IUD after year of usage of the IUD. There are three collagen vacuoles present, but in the immediate surroundings of the cell no collagen is visible, ×5700

replacement with a new IUD is recommended after 12 months, it is quite common for patients not to return for replacement until after 14 or 15 months. Endometrial biopsies from these patients as well as biopsies or hysterectomies from patients who have used consecutive IUDs for periods of up to 4 years reveal a morphology very much different from that described above.

With prolonged usage of the progesterone-releasing IUD the organelles involved in synthesis, namely the ER and the Golgi apparatus, are reduced drastically. The decidual granules have disappeared almost totally and the number of collagen vacuoles is also diminished. The ones that remain possess hardly any collagen fibrils, but are more or less electron lucent (Fig. 54). The BL is not reduced to any extent, even after several years of usage of the IUD. Miniature mitochondria are still present, though slightly reduced in number.

3.3.7 Glandular and Surface Epithelium

Even in the first few months following insertion of the IUD, a pronounced effect of the progesterone exerted on the glandular and surface epithelium can be detected. The epithelium does not respond to the increase in endogenous hormone during the secretory phase of the cycle, and the postovulatory triad is not formed. The height of the epithelium remains basically unaltered throught-out the cycle and glycogen is not released from the epithelial cells into the

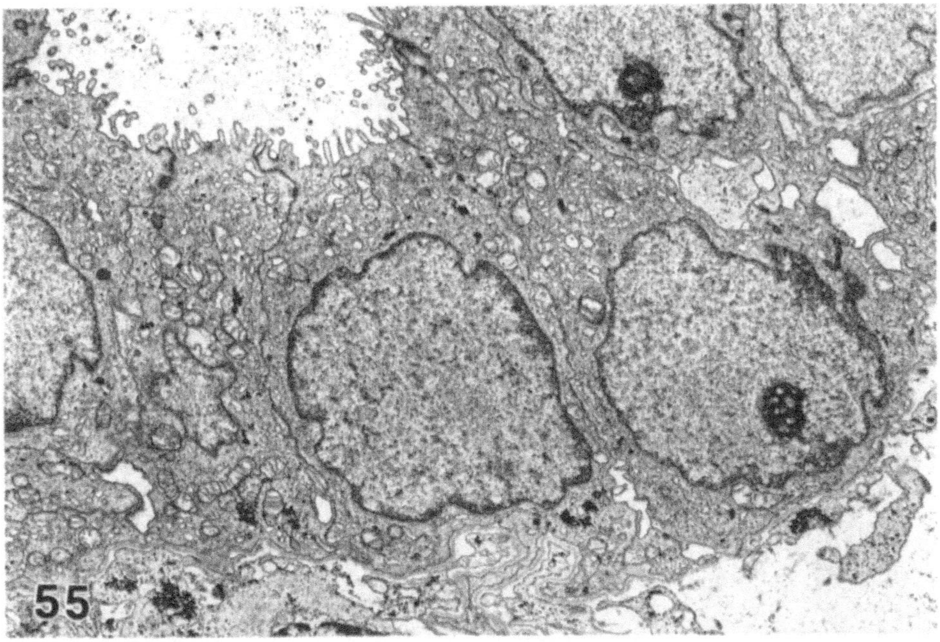

Fig. 55. Glandular epithelial cells after 1 year of usage of a progesterone IUD. The biopsy was taken during the middle of the secretory phase. The cells are cuboid and appear to be almost atrophic, ×9600

glandular lumen (Fig. 55). During later months, the cells maintain more or less the same kind of morphology, except in their subnuclear regions where they start to store lipid droplets instead of glycogen. With continued usage of the IUD, the cells reduce their activities even more and acquire a rather atrophic morphology (Fig. 55). The surface epithelium remains unusually flat during the course of a cycle. The above-described changes also seem to be dependent upon the local progesterone concentration. Thus, the closer the epithelium to the IUD, the more pronounced the changes. However, even after several years of usage of the IUD a certain heterogeneity of the response to the exogenous progesterone was always observed.

3.4 The Endometrium Under a Copper IUD

3.4.1 The Surface Epithelium (SEM)

Although the effects of copper IUDs on the surface epithelium are clear cut and easily detectable, the physiological variations of the menstrual cycle always have a bearing on the effects of the copper.

The effects exerted by the IUD on the surface epithelium can be divided into two groups: (a) effects in areas in direct contact with the IUD, and (b) effects in areas distant from the IUD.

The effects in areas in direct contact with the IUD are the most obvious ones. In most cases an imprint can already be detected with the naked eye after

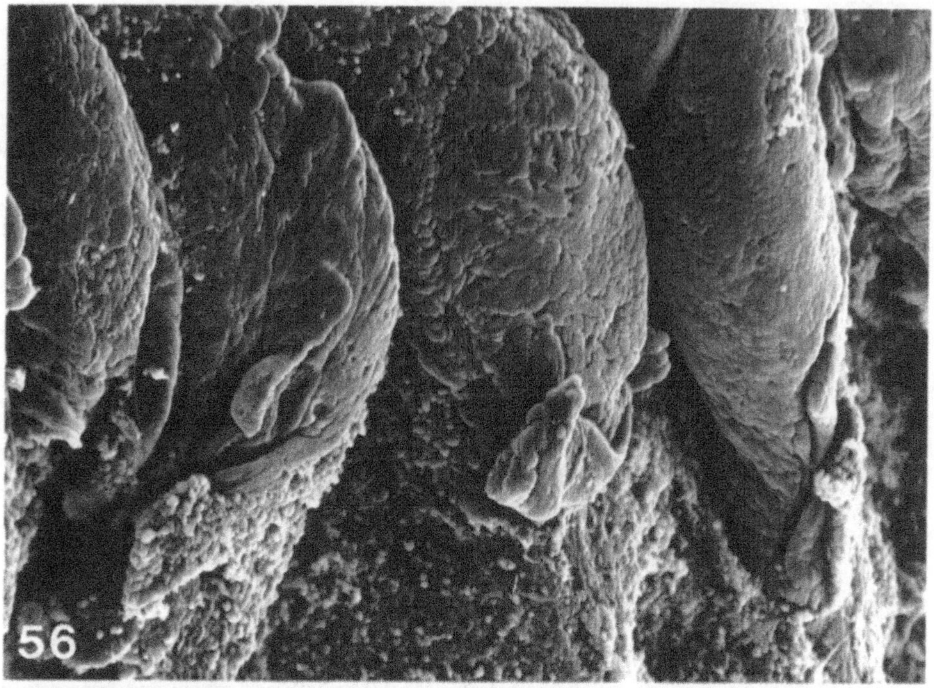

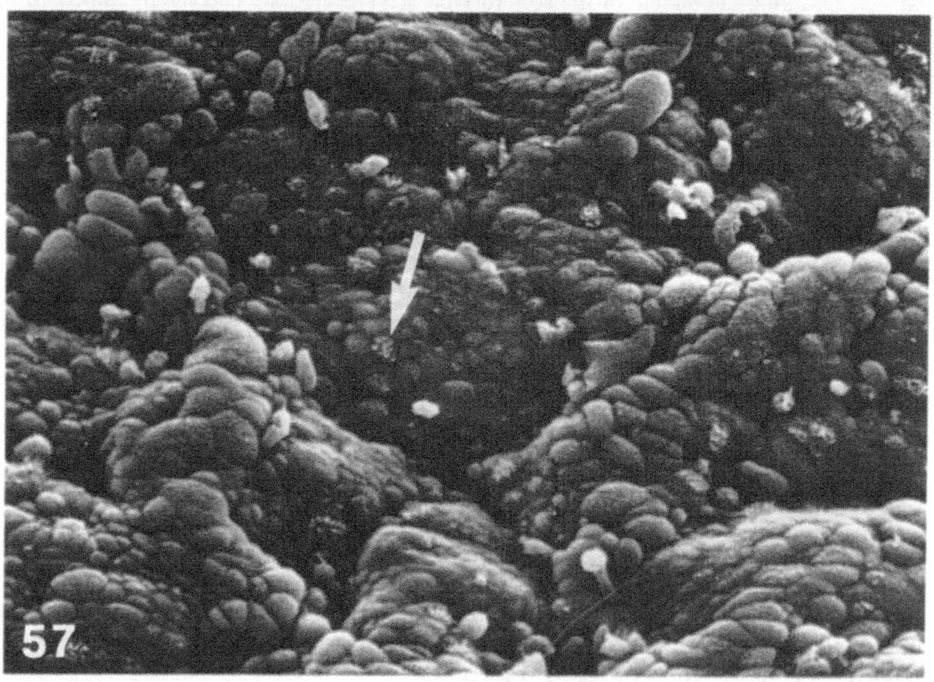

Figs. 56, 57. SEM photographs from endometrium under a copper IUD **Fig. 56.** Deep impressions caused by the copper wire from the IUD. In the *lower third* the surface epithelium is damaged and the stroma is exposed, ×100. **Fig. 57.** Typical picture of IUD contact area. The surface epithelium is rather heterogeneous and the number of ciliary cells is drastically reduced (*arrow*). Middle of the secretory phase, ×490 removal of the IUD (Fig. 56). The pattern of the imprint corresponds well to the copper wire which is wound around the vertical stem of the IUD. Inert IUDs simply produce a flat imprint without a pattern. In the areas of the imprint the surface epithelium is much flatter than in other areas of the uterine cavity. Due to mechanical wear and probably also due to the copper ions released from the IUD the cell membranes are relatively fragile. This is reflected by the frequent tissue damage present in apical cell regions which might well be caused through preparative procedures, and in focal areas where the whole epithelium including the BL is missing. At sites of such severe damage the underlying stroma is exposed (Fig. 56) – this is only found in tissue areas which are in direct contact with the IUD.

Another typical feature of the zone of direct contact is a decrease in the number of ciliated cells, throughout the whole menstrual cycle there are only a very small number of ciliated cells present (Fig. 57). The nonciliated cells have lost most of their microvilli and the few which are still present resemble those from postmenopausal women (cf. Ludwig and Metzger 1976a). In contact areas as well as in areas distant to the IUD the microvilli have changed their morphology. They are shorter than normal (Fig. 58). Many cells have developed microridges (Fig. 61). Intermediate stages between microvilli and microridges can also be found. The surface epithelium has acquired a heterogeneous appearance; the shape of individual cells is not as uniform as is the case in the normally cycling endometrium (Fig. 62).

The most outstanding feature of the endometrium under a copper IUD is the frequent formation of micropolyps, which can already be seen in light microscopic sections. These micropolyps are mainly present in the direct contact area and immediately surrounding area of the IUD (Figs. 63–65), and they occur both as singular micropolyps (Figs. 64, 65) and in groups of several micropolyps (Fig. 63). Their surface epithelium shows a very high degree of heterogeneity (Fig. 65); minute cells are located right next to giant cells. The microvilli of the cells covering the micropolyps are either unusually short or not present at all (Fig. 65). Ciliated cells are never present on the micropolyps, not even when groups of ciliated cells are found near to their site of origin. The micropolyps are always vascularized; sometimes several vessels of various sizes can be found. As a result of the FBR to the IUD, the endometrium is also infiltrated by high numbers of leukocytes, although most of those which have reached the luminal side of the endometrium are probably washed away during preparation of the tissue for SEM. The leukocytes that adhere to the IUD for the most part endure the preparatory procedures without any damage and occupy most of the inert plastic parts. A prerequisite for this adherence appears to be a surface coat on the IUD consisting of cellular debris and uterine fluids (see Spornitz et al. 1984). On absolutely clean plastic areas, leukocytes are hardly ever found. Occasionally mucous was found covering the surface epithelium. This was mostly the case in areas adjoining the uterine cervix, although it was also observed in the corpus and fundus regions of uteri in patients with dysfunctional bleeding.

The effects in areas not in contact with the IUD depend upon the distance at which they are located from the IUD. The further away from the IUD, the less pronounced the effects. In the upper fundus region and the area of the

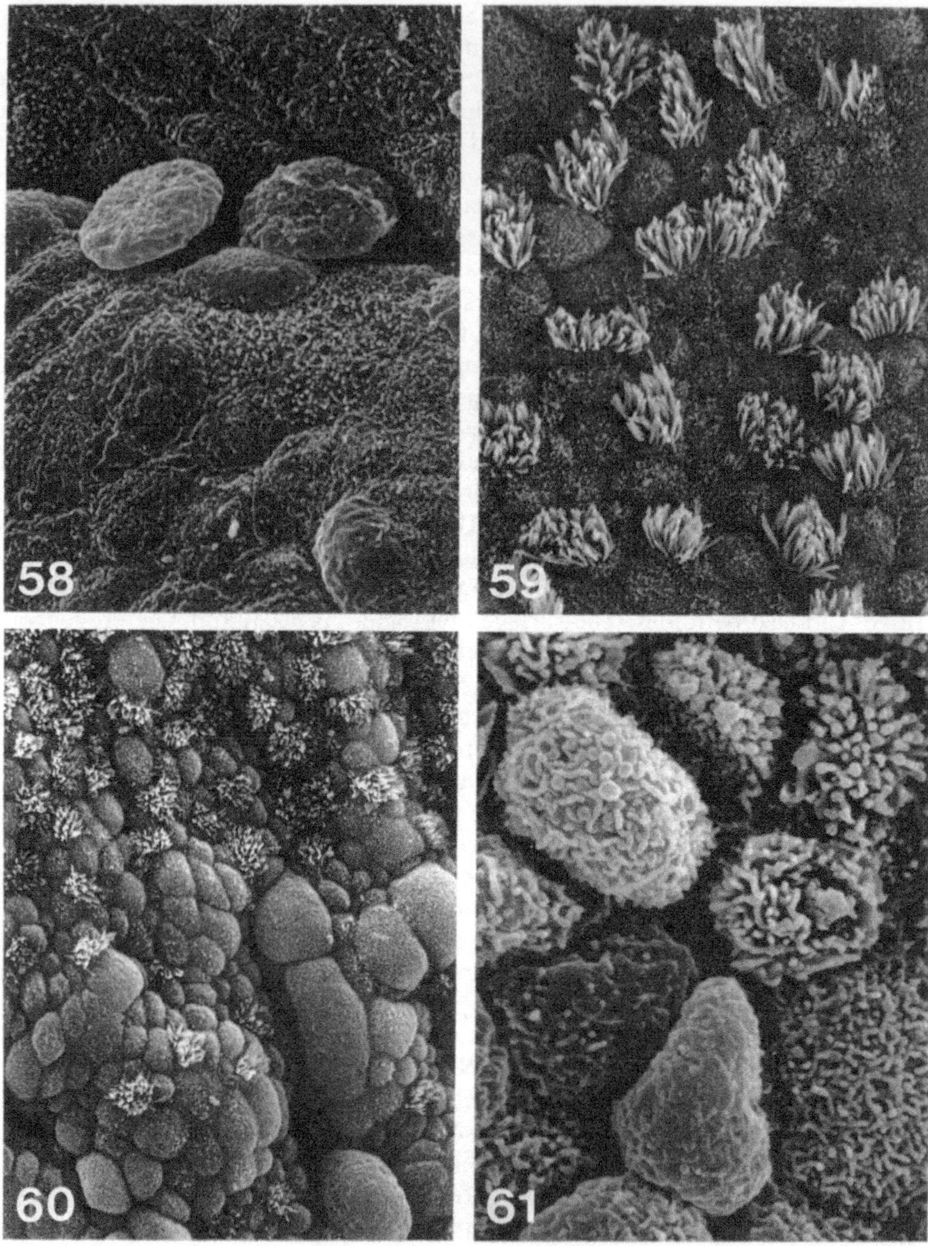

Figs. 58–61. Various aspects of surface epithelium under a Cu-IUD. **Fig. 58.** Surface epithelium in direct contact with the IUD. The microvilli are reduced in number and in height. Ciliary cells are totally absent. The epithelium resembles that from the senium ×1800. **Fig. 59.** Surface epithelium from the area distant to the copper IUD; the morphology corresponds well to the phase of the cycle (middle of the secretory phase). Ciliary cells are unaltered, ×1250. **Fig. 60.** At the *top* the epithelium appears normal but at the *bottom* the influence of the copper released from the IUD is clearly visible, ×550. **Fig. 61.** Heterogeneity of surface epithelium as it is frequently found under a copper IUD. There are cells without microvilli next to cells with normal or altered microvilli. In contrast to untreated endometrium even microridges are formed, ×3200

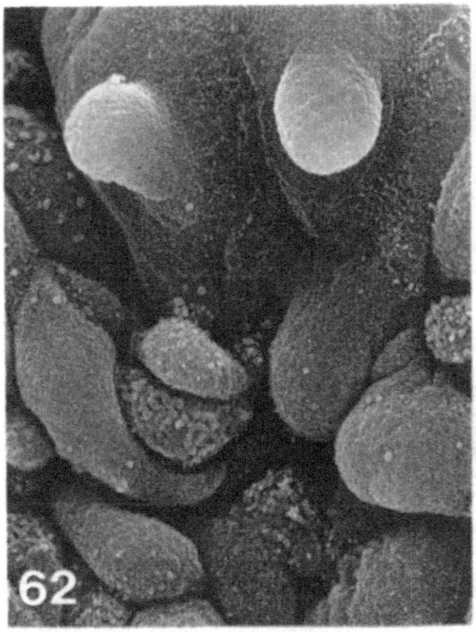

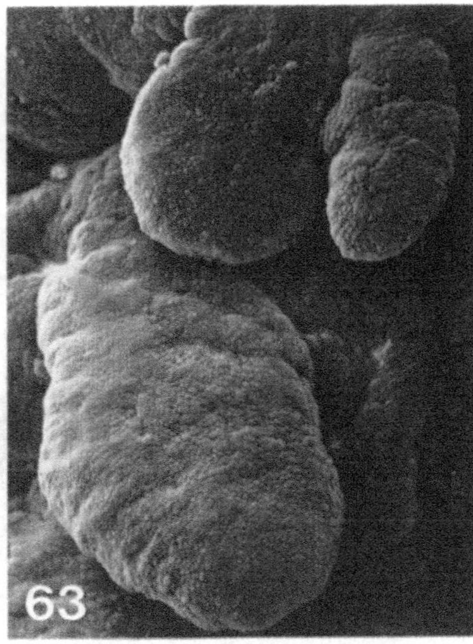

Figs. 62, 63. Surface epithelium under a copper IUD. **Fig. 62.** Heterogeneity typical for area in direct contact with IUD is clearly visible. This might be the initial stage of micropolyp formation, ×2400. **Fig. 63.** Micropolyps, characteristic of area in direct contact with the IUD; they are restricted to this area, ×135

tubal openings hardly any changes are detected. Micropolyps are not formed in the distant areas. Frequently a relatively sharp demarcation line marks the limit to which ciliated cells grow. Nearer to the IUD, no ciliated cells can be found (Fig. 60), while further away from the IUD numerous ciliated cells are present (Fig. 59).

3.4.2 The Stroma and the Glandular Epithelium (TEM)

The action of copper IUDs on the glandular epithelium is not as pronounced as the action on the surface epithelium. The glandular epithelium of the endometrium which is in direct contact with the IUD does, however, have a reduced synthesis of glycogen during the secretory phase of the cycle. Since, in all other tissue areas, the postovulatory triad, namely glycogen, giant mitochondria, and the NCS is formed to an extent comparable to that of untreated endometrium, the glandular epithelium under copper IUDs can be used for dating the endometrium.

The stroma, on the other hand, changes considerably due to the action of the IUD. Mechanical stimulation of the endometrium, such as that caused by copper IUDs, may, for example, very likely induce decidualization. This is however, only the case in the tissue area in direct contact with the IUD. The lateral areas of the endometrium, i.e., areas which are not in contact with the IUD, the more profound areas of the stratum functionale, and the whole of the

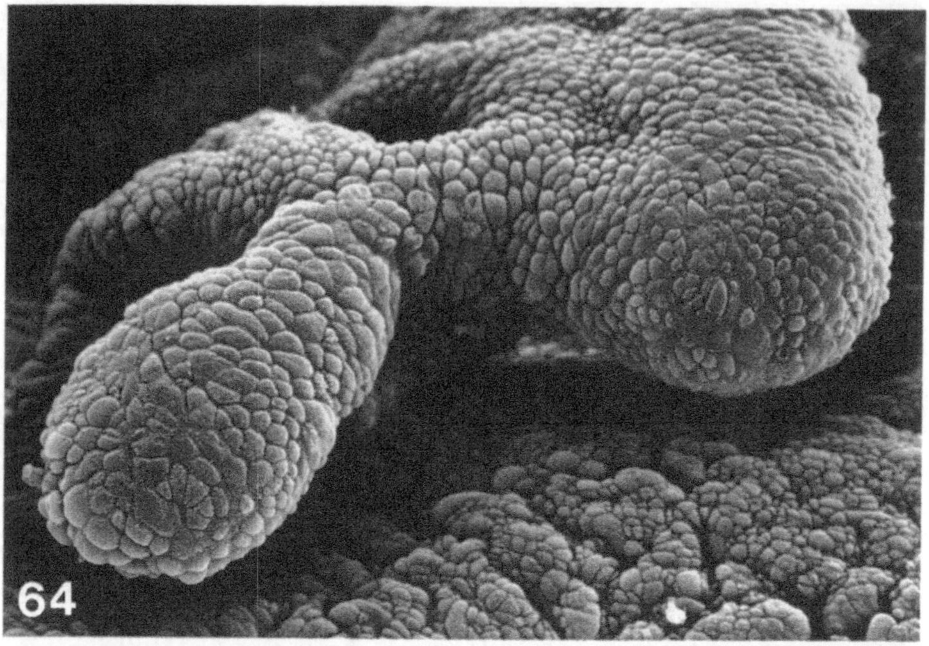

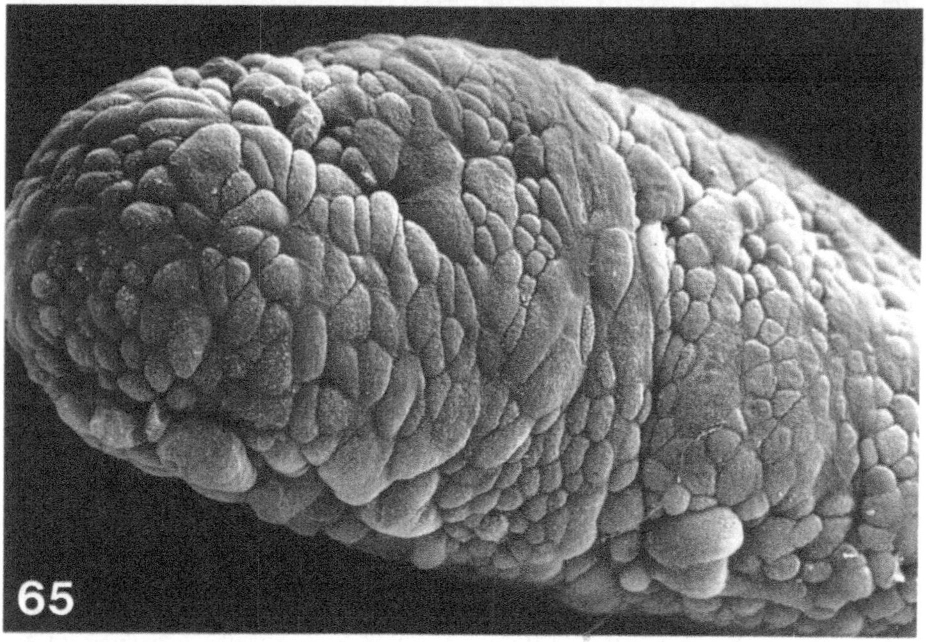

Figs. 64, 65. Micropolyps from the contact area of a copper IUD. **Fig. 64.** Branched micropolyp. The surface epithelium does not carry any ciliary cells, ×380. **Fig. 65.** Detail of surface epithelium from another micropolyp. The heterogeneity of the epithelium is clearly visible. There are hardly any microvilli which can be taken as evidence that this epithelium was in direct contact with the IUD, ×560 stratum basale are hardly ever decidualized through the action of a copper IUD. The most obvious change in the stroma, however, is the FBR which is reflected by the presence of numerous leukocytes. Where there is a pronounced FBR, infiltrating leukocytes are found not only in the stroma, but also in the uterine cavity and in the lumen of the endometrial glands. Copper IUDs which have been in the uterine cavity for more than 1 year are usually covered on their inert plastic parts by thousands of leukocytes. Most of the leukocytes are neutrophilic granulocytes. In severe cases of FBR lymphocytes are frequently found, sometimes even forming lymph follicles.

3.4.3 The Decidua Under a Copper IUD

The decidualization of the stroma under copper IUDs is usually a focal one. Two mechanisms are apparently responsible for focal decidualization, which can be inferred from the localization of the decidual foci. The most frequently occurring decidual foci are the superficial ones which, as previously stated most likely result from the mechanical trauma caused by the IUD. The cells in these focal points are only partially decidualized. They have already reduced the volume of their mitochondria and their cell volume has been greatly enlarged, but they do not, however, possess decidual granules or a continuous BL. The second type of decidualization is likewise only a partial one, also lacking decidual granules and a BL. It occurs in deeper regions of the endometrium and appears to be dependent upon the actual phase of the cycle, i.e., it occurs only during the late secretory phase of the cycle. This type of decidualization initially takes the form of periarterial sheathing of stromal cells. Thus, periarterial coats of decidualized cells are formed which occasionally grow to such an extent that they come into contact with the more superficial decidual foci. Both types of partial decidualization, i.e., the superficial and the periarterial decidualization are characterized by the presence of collagen vacuoles, similar to those found during the formation of late predecidual cells. With superficial decidual foci, a desynchronization of glandular and stromal development is frequently the case, leading to the presence of subnuclear glycogen in the glandular cells and the simultaneous presence of predecidual cells in the stroma.

Partial decidualization caused by copper IUDs appears to be independent of the type of IUD used, as it was found with all types of IUDs investigated in this study.

3.5 The Decidua of Pregnancy

3.5.1 General Aspects

Deciduomata such as those caused by the use of copper or progesterone IUDs can only be classified with respect to the degree of decidualization. With the true decidua as it is formed during pregnancy there is a clear-cut differentiation

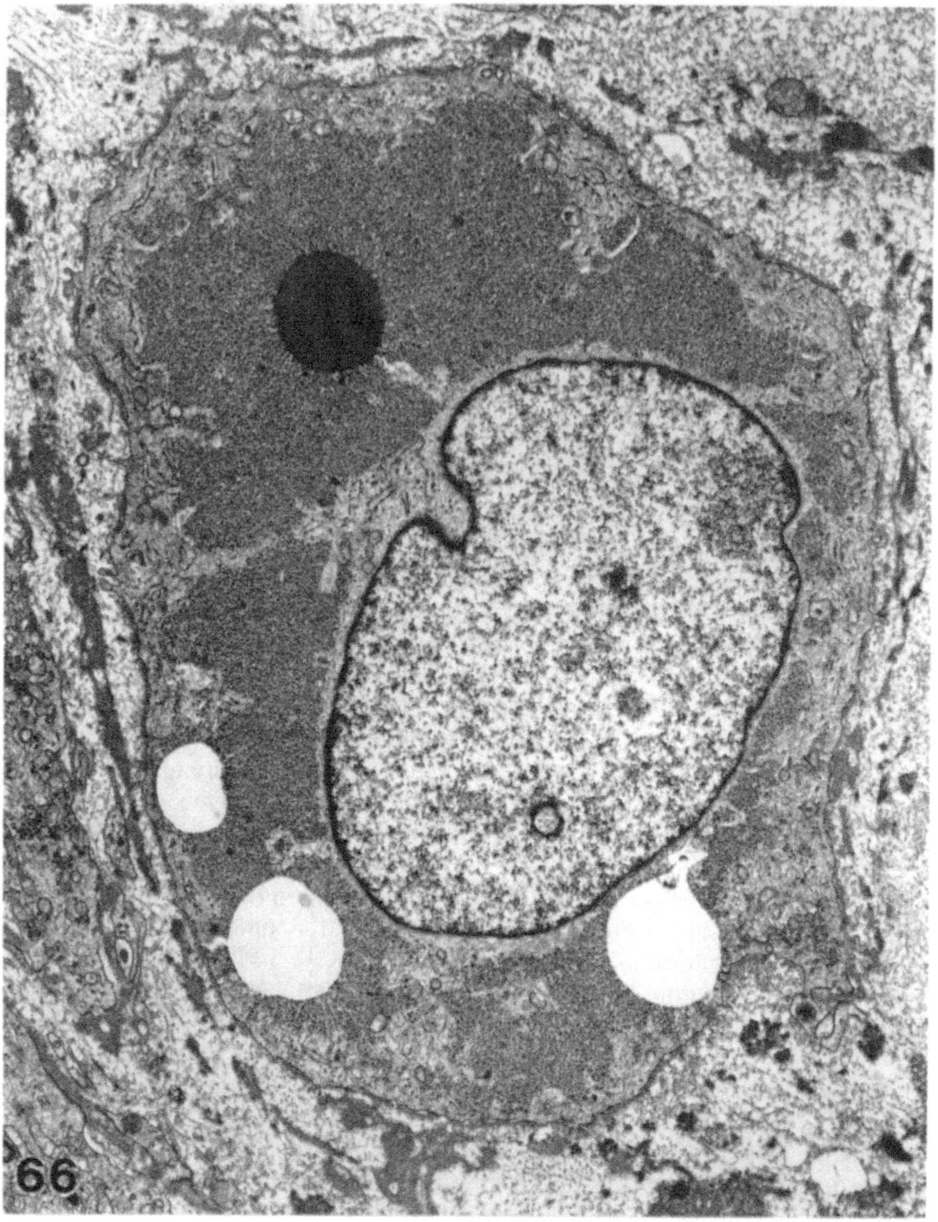

Fig. 66. Decidual cell from the decidua basalis (3rd month of pregnancy). These cells are usually filled with glycogen and lipid droplets. They are studded with only a few decidual granules, ×8000

with respect to the localization of the decidual tissue which makes classification of the different areas a necessity. Thus, three zones are usually distinguished:

1. Decidua basalis (underlying the implantation site).
2. Decidual parietalis (covering the walls of the uterine cavity).
3. Decidua capsularis (covering the implantation site).

The decidua capsularis degenerates and the underlying chorion fuses with the decidua parietalis around the 5th month of pregnancy (Hamilton et al. 1976). The decidua basalis is of the greatest interest, since it is formed in the area of the primary uteroplacental interface and plays an important role during the whole process of implantation. Furthermore it is the site where immunologic processes between mother and child take place.

The decidua basalis mostly comprises round decidual cells, while the decidua parietalis and the region between the decidua parietalis and decidua basalis (formerly called decidua marginalis) mainly comprise spindle-shaped decidual cells.

3.5.2 Decidua Basalis

The cells of the decidua basalis are the first to be developed during the formation of a genuine decidua of pregnancy. They are differentiated from the predecidual cells of the late secretory phase and their morphology is that of typical decidual cells – they are round, relatively large cells with a BL and varying amounts of lipid and glycogen (Fig. 66). They are usually surrounded by fibrinoid which constitutes part of the uteroplacental interface. Their development includes the formation of decidual granules, though not as many as the decidual cells induced as a result of the presence of a progesterone-releasing IUD produce. The decidual cells of the decidua basalis do, however, possess all the typical features of decidual cells, such as miniature mitochondria. Compared with the progesterone IUD-induced decidual cells, the nucleocytoplasmic ratio has shifted to the side of the cytoplasm (see Sect. 3.6). In the areas where large amounts of fibrinoid are formed, decidual cells frequently degenerate and are subsequently integrated into the newly formed fibrinoid. The decidual cells in these areas have lost all of their decidual granules and are surrounded by masses of fibrinoid which, unlike the fibrinoid formed through the action of progesterone IUDs, is intermingled with collagen. The cell membranes of cells in these areas also have numerous undulations which are filled with BL. The BL is continuous with the surrounding fibrinoid. Occasionally cellular detritus is produced which is apparently incorporated into the fibrinoid. In the area of the decidua basalis fetal cells may be encountered. These are mostly cytotrophoblast cells which have migrated into the tissue from the columnar trophoblast of the anchoring villi. They can easily be distinguished from decidual cells by their normal-sized mitochondria and the absence of a BL coat.

3.5.3 Decidua Parietalis

The decidua parietalis has a morphology which is much looser, or sponge like than that of the decidua basalis (Figs. 67–69). The distance between the individual decidual cells is relatively great and there are very few collagen fibrils in the intercellular space (Fig. 69). The decidual cells are round to oval and are similar in size to the cells of the decidua basalis. They contain only a

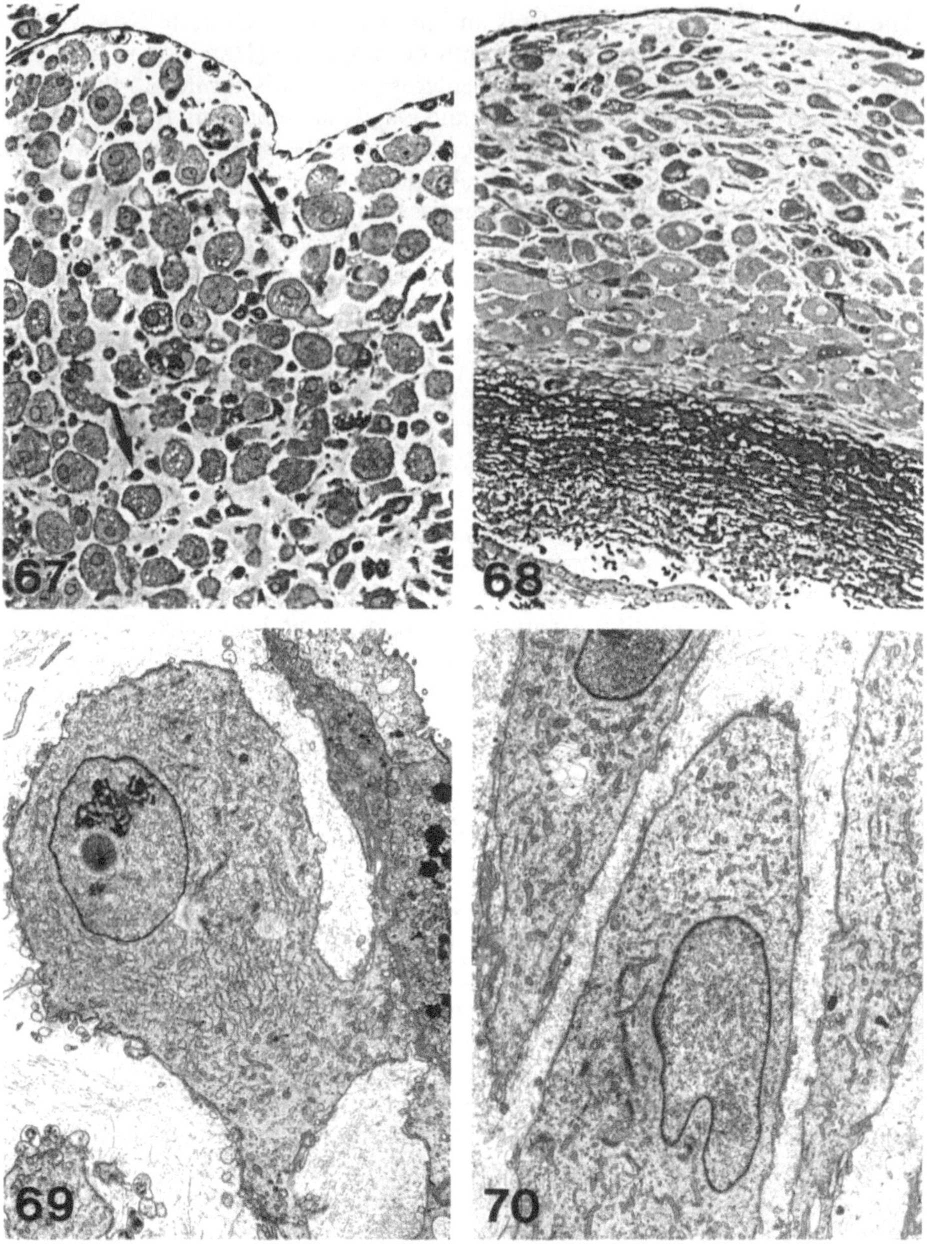

Figs. 67–70. Decidua parietalis and decidua capsularis. **Fig. 67.** Semi-thin section from decidua parietalis (3rd month of pregnancy). The surface epithelium is extremely thin. There is very little fibrinoid. K cells are still present, though reduced in size (*arrows*), ×225. **Fig. 68.** Semi-thin section from decidua capsularis (3rd month of pregnancy). The decidual cells are spindle shaped in the subsurface area and tend to become round towards the broad layer of fibrinoid (at the *bottom*), ×225. **Fig. 69.** Decidual cell from decidua parietalis in direct contact with the extremely thin surface epithelium. The intercellular space is free from fibrinoid. There are very few decidual granules, ×3700. **Fig. 70.** Decidual cells from decidua capsularis. The cells are spindle shaped and lie parallel to each other. They are longer than cells from both decidua basalis and parietalis, ×5000

limited number of decidual granules and very little glycogen; they do, however, have numerous lipid droplets. A typical BL coat and miniature mitochondria are also present. The mitochondria of the decidua parietalis contain mitochondrial cristae of the tubular type, which are more in evidence here than in other decidual areas.

3.5.4 Decidua Capsularis

The decidua capsularis is very different in appearance to the decidua basalis, largely because (a) the cells are mostly spindle shaped, and (b) there are hardly any fetal cells present. The decidual cells have a longitudinal diameter of 40–45 μm and are about 8–10 μm wide. Owing to their spindle-like shape, the cells are mostly oriented parallel to each other (Fig. 70) The BL coat is only about 50 nm thick. Decidual granules are rarely present in these cells, but the synthetic apparatus is comparable to that of cells from the decidua basalis. The intercellular space is wide and there is only very little fibrinoid present. The few collagen fibrils present run parallel to the long axis of the cells. Towards the trophoblast there is, however, a large amount of fibrinoid (Fig. 68) which contains numerous degenerating decidual cells and blood cells which are apparently incorporated into the developing fibrinoid. The degenerating decidual cells have a unique feature, inasmuch as their mitochondria develop an electron dense matrix, the development of which is accompanied by the loss of mitochondrial cristae. In this way, electron dense particles (mitochondrial granules) are formed which are bounded by a normal mitochondrial double membrane. Since this only occurs in areas where decidual cells are incorporated into the fibrinoid, i.e., in areas close to the trophoblast, it seems very likely that the appearance of mitochondrial granules can be taken as an early indication of degeneration. Towards the trophoblast the decidual cells apparently change their shape and develop a rounded profile (Fig. 70).

3.6 Morphometry of Decidual Cells

3.6.1 Decidua of Pregnancy

In order to provide a quantitative basis for the comparison of decidual cells from different origins, e.g., ovarian decidual cells, progesterone IUD-induced decidual cells, and cells from normal pregnancy, a morphometric study was carried out. Among the parameters measured were the volume occupied by the synthetic apparatus, the amount of paraplasm, and the organelle-free part of the cytoplasm, which all serve as a criterion for the activity of the cell. Both the absolute size of the cell and the volume density (VV) of the individual cellular compartments were measured. The cells of the decidua basalis were taken as representative of the decidua of pregnancy, since the cells from the decidua parietalis and capsularis represent only a small portion of the total number of decidual cells.

On the basis of their total volume the cells from the decidua basalis can be divided into two groups: large cells and small cells. When cut through the region of their largest diameter the section area of the large cells measures 730 μm^2 whereas the section area of the small cells measures only 210 μm^2. The calculated average section area of all cells measures about 420 μm^2.

The (VV) of individual organelles was calculated as a percent age of the cell volume (Table 2).

When comparing small decidual cells to large ones it becomes obvious that the composition of organelles varies considerably in the two types of cells. The greatest difference between large and small decidual cells lies in the composition of the synthetic apparatus, i.e., the Golgi apparatus and the ER. The ER/Golgi ratio of the small cells amounts to 2.32, whereas the ER/Golgi ratio of the large cells amounts to 10.82.

The difference in the VV of the nucleus is only partly due to the growth of the cells; it can also be attributed to the growth of the nuclei themselves as they apparently increase their own volume. The nuclei of large cells are approximately 10% larger than the nuclei of small cells.

3.6.2 Progesterone IUD-Induced Decidua

When cut through the region of their largest diameter the progesterone-induced decidual cells have a section area of about 500 μm^2. Unlike the decidual cells from pregnancy, the IUD-induced decidual cells cannot be divided into two subpopulations as they consist of a gaussian population which ranges from the smallest to the largest cells. The VV of the individual organelles calculated as a percentage of the cell volume is shown in Table 3.

The ER/Golgi ratio of the progesterone-induced decidual cells amounts to 2.41, which is almost as low as that of the small decidual cells from pregnancy. Of particular note is the high figure for decidual granules (4.38%) which is about three times as high as that for the decidual granules in decidual cells from pregnancy.

3.7 Decidua of Ectopic Pregnancy

Three cases of ectopic pregnancy, where the blastocyst had implanted in the fallopian tubes, were investigated. On the basis of light microscopic sections, it was expected to find genuine decidual cells, either of endometriotic origin or induced from tubal fibroblasts. Electron microscopic studies revealed, however, that in none of the investigated cases were there any genuine decidual cells present. The relatively large cells present, which on light microscopic sections looked similar to decidual cells, did not have any of the features typical of decidual cells, except their size. They did not have miniature mitochondria, decidual granules, or a BL. Therefore, it is suggested these cells should be named "pseudodecidual cells". They are of "columnar" cytotrophoblastic origin and invade the tissue at the implantation site via anchoring villi, which are particularly numerous in tubal pregnancy. Unlike in eutopic

Table 2. Volume density of individual organelles as a percentage of cell volume (Decidua from I. term of pregnancy)

Organelle	Cell volume (%)		
	Average decidual cell	Small decidual cell	Large decidual cell
Mitochondria	7.73	7.54	7.42
Golgi apparatus	2.05	2.90	1.03
ER[a]	8.32	6.73	11.14
Decidual granules	1.63	1.03	1.71
Paraplasm	12.97	12.23	13.95
Remaining cytoplasm	55.17	54.50	55.07
Nucleus	12.14	15.06	9.69

[a] ER, endoplasmic reticulum

Table 3. Volume density of individual organelles as a percentage of cell volume (Progesterone IUD-induced decidua)

Organelle	Cell volume (%)
Mitochondria	7.08
Golgi apparatus	3.17
ER[a]	7.65
Decidual granules	4.83
Paraplasm	4.37
Remaining cytoplasm	58.72
Nucleus	14.16

[a] ER, endoplasmic reticulum

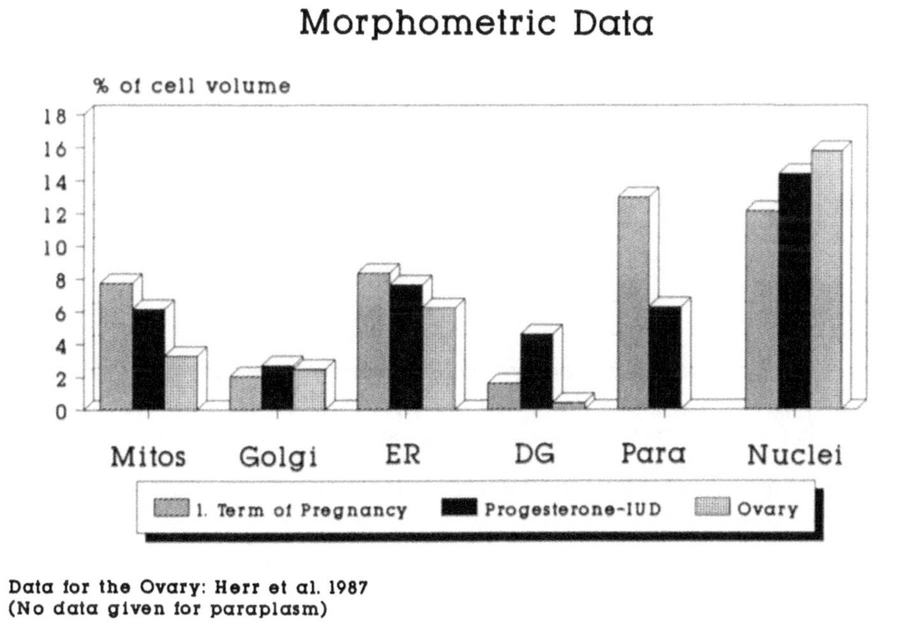

Graphic 1. Decidual cells

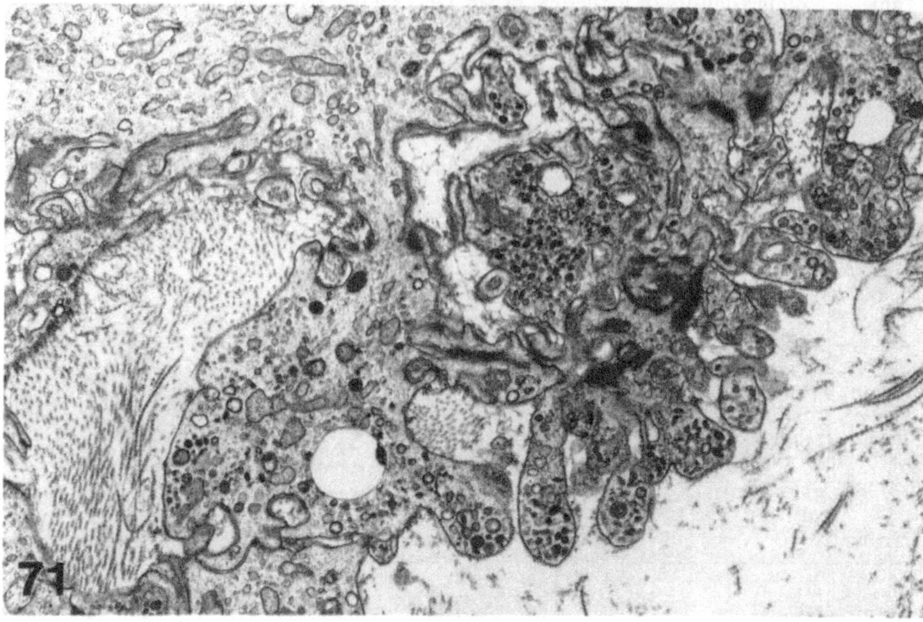

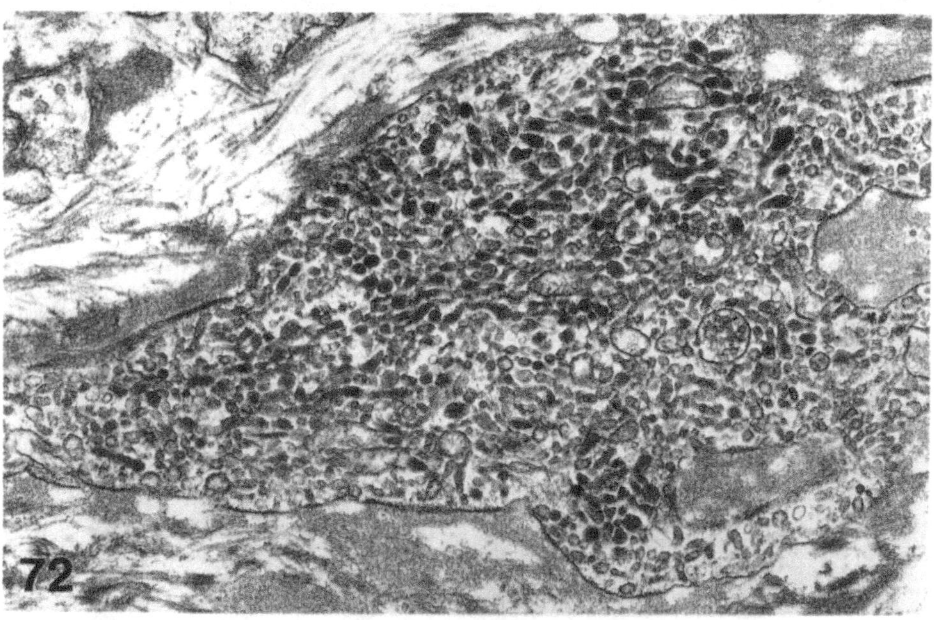

Figs. 71, 72. Secretory granules in decidual cells. **Fig. 71.** Numerous secretory granules present in finger-like extensions of the decidual cells. The cellular extensions are connected to the cell by means of a narrow bridge of cytoplasm, ×13 000. **Fig. 72.** Extension of decidual cell from 3rd month of pregnancy filled with secretory granules to such an extent that hardly any other organelles can be seen. The BL coat is evidence of the decidual nature of this cell, ×20 000 pregnancy there is practically no fibrinoid formed at the implantation site, this is a good indication of the decidual origin of fibrinoid. An ultrastructural study on ectopic pseudodecidual cells will be published elsewhere.

3.8 Secretory Granules of Decidual Cells

The decidual cells at all stages of maturation contain a second type of secretory granule which frequently escapes attention because of its specific localization within cellular spiky extensions of the decidual cells (Figs. 71, 72). These extensions are even present in mature more or less round, decidual cells from the decidua basalis. On the average section through a decidual cell, the thin connection of the extensions which contains the secretory granules is hardly ever cut, although serial sectioning shows that the cellular extensions are connected to the cells by a narrow bridge of cytoplasm. The secretory granules in these areas of cytoplasm are very heterogeneous with respect to both their size and morphology. They are surrounded by a trilaminar membrane similar to that present in the Golgi apparatus. The majority of these granules are round to oval and measure between 30 and 350 nm. Many, however, are oblong (Fig. 73), the longest being up to 700 nm in length and about 80 nm in width. Some granules are filled with an electron dense, fine granular content, while others are electron lucent – making them appear empty. Towards the end of the first trimester of pregnancy or after about 10 months of usage of the progesterone IUD, there appears to be a maximum of these granules in the decidual cells. More than a thousand granules can be counted on an average section through the granule-containing regions of the cells. Even when the cells no longer possess any decidual granules, decidual cells can always be unequivocally identified as secretory granule-containing cells on the basis of the BL covering them; other cells such as K cells, or undifferentiated fibroblasts are never covered by a BL. A characteristic feature of the secretory granule-containing decidual cells is the presence of secondary protrusions from the spiky extensions of the cells (Fig. 74). These secondary protrusions are, however, only found when secretory granules are present; other cellular extensions which do not contain secretory granules never have secondary protrusions. As is the case with other active secretory cells, the actual process of secretion can only in rare instances be seen on TEM. Discharged secretory granules are, nonetheless, frequently found in the intercellular space neighboring the decidual cells (Fig. 73). They appear to be discharged from the cells by an apocrine mechanism, as they are still surrounded by their limiting membrane. Another mechanism of discharge appears to be the detaching of primary and/or secondary cellular extensions which occasionally are found in areas remote from any decidual cell. Digestion, as done on ultra-thin sections, with α-amylase, and lipase, does not alter the morphology of the secretory granules in any way. After treatment with pronase, a rather unspecific proteolytic enzyme, the contents of the secretory granules vanish almost totally from the sections.

A topographical relationship between the secretory granules and the Golgi apparatus was in no case established. There are, however, always

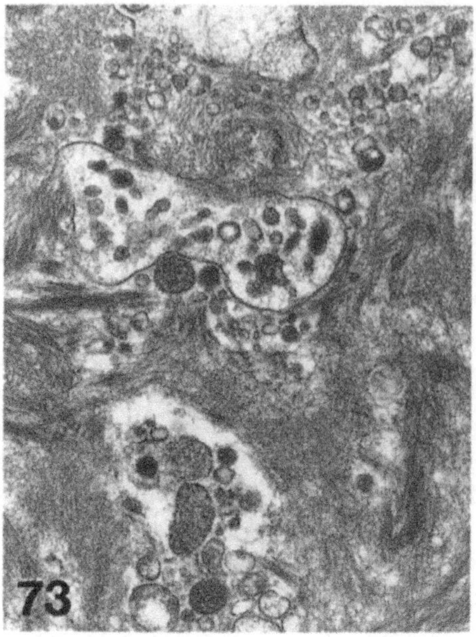

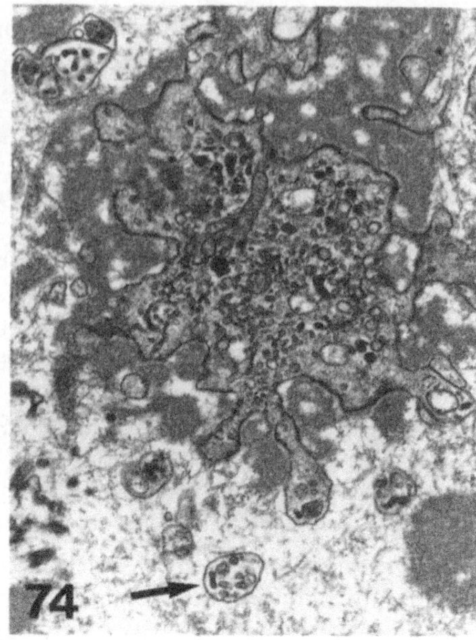

Figs. 73, 74. Secretory granules in decidual cells. **Fig. 73.** Extension of a decidual cell amidst fibrinoid. The extension is filled with secretory granules. Outside the extension numerous secretory granules are present which apparently have been discharged from the cell, ×22 000. **Fig. 74.** Primary extension from a decidual cell filled with secretory granules. Numerous secondary extensions are visible. The primary extension has a BL coat, ×13 500

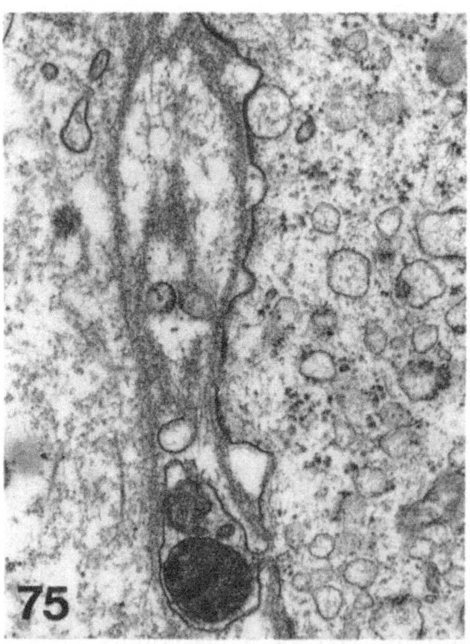

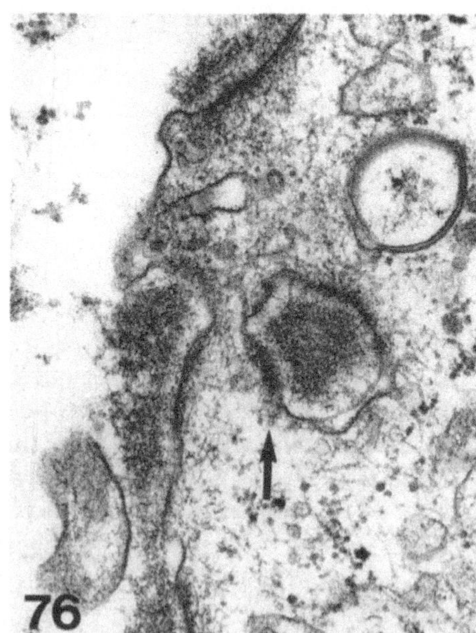

Figs. 75, 76. BL coat on decidual cells. **Fig. 75.** Smaller undulations of cell membrane (*right*) are not lined with BL, ×8000. **Fig. 76.** Larger undulations or invaginations are regularly lined by BL (*arrow*), ×22 000

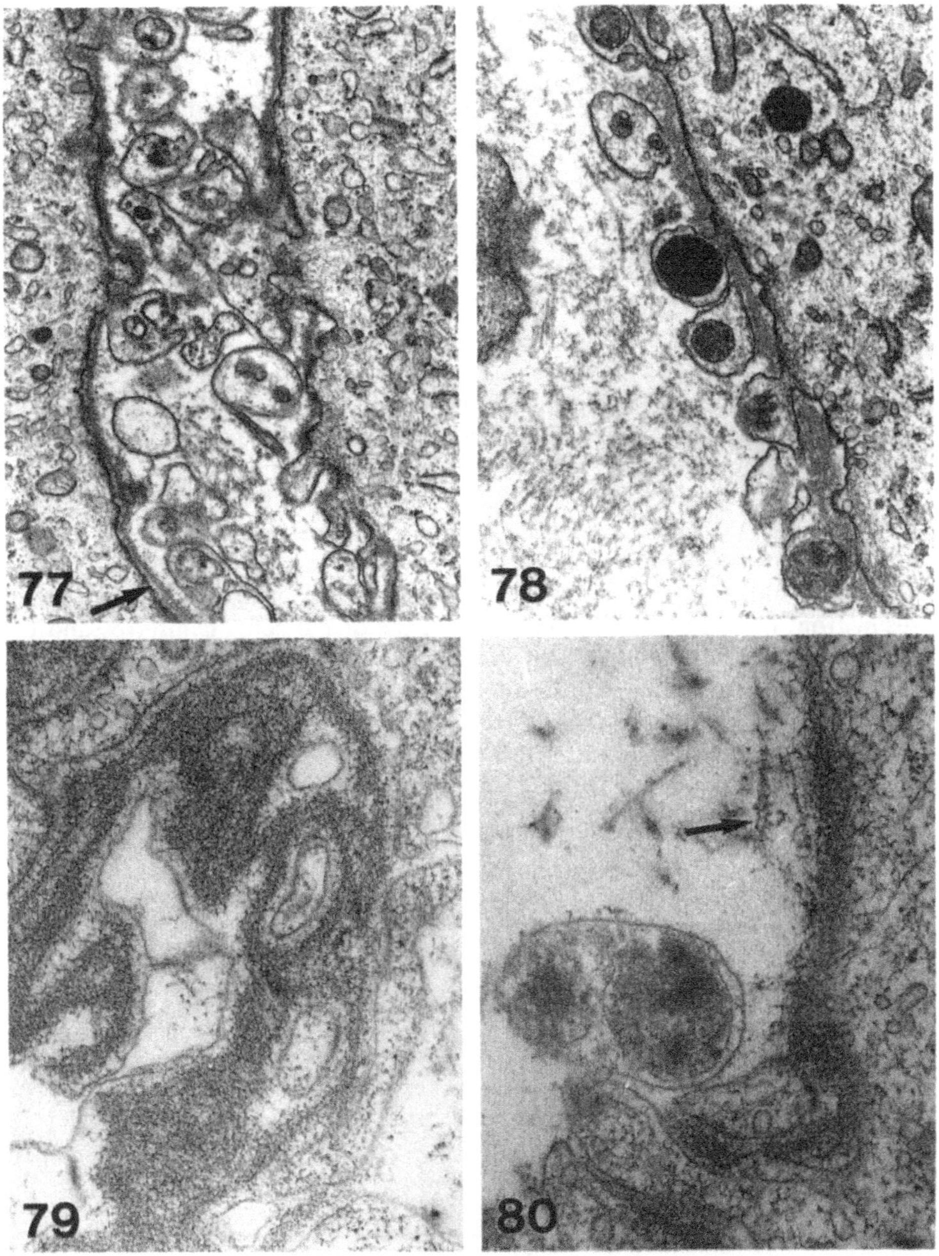

Figs. 77–80. BL coat on decidual cells. **Fig. 77.** Progesterone IUD-induced decidual cells with BL. The BL of these cells has a distinct lamina rara interna (*arrow*) which is about 40 nm wide, ×18000. **Fig. 78.** Decidual cell from pregnancy. The BL is wider and possesses hardly any lamina rara interna, ×19000. **Fig. 79.** BL on a decidual cell from the 3rd month of pregnancy after digestion with α-amylase. Now a distinct lamina rara interna, which is about 50 nm wide, is present, ×35000. **Fig. 80.** After short digestion with pronase the connection of the BL with anchoring fibrils (*arrow*) becomes visible, ×37000

morphologically identical granules present in the Golgi areas. These granules unfortunately resemble the precursors of the decidual granules which makes positive identification almost impossible.

3.9 BL of Decidual Cells

Decidual cells are always covered by a continuous BL, regardless of the place of origin (ectopic or eutopic) or the mechanism of their induction (pregnancy or progesterone IUD; Figs. 75, 76). In cases where decidual cells are located close to either surface or glandular epithelium, the BL of decidual cells is continuous with the BL of these cells. The BL induced through pregnancy is different from the BL induced through a progesterone IUD (Fig. 77). The average BL from pregnancy measures about 420 nm, i.e., 0.46 μm, in width (Fig. 78); it does, however, vary considerably in thickness. Thus, on one and the same cell the thickness can easily range from 100 nm (0.1 μm) to over 1000 nm (1 μm). The BL induced through progesterone IUDs is on average only 100 nm wide, in a range extending between 70 and 200 nm. It usually does not follow all the many smaller undulations of cell membranes which might result from the different amount of shrinkage of the cells themselves and the BL. Larger undulations, like invaginations, are usually lined by BL.

After digestion with *α*-amylase the BL shows a distinct lamina rara interna which is about 50 nm wide (Fig. 79). After digestion with pronase a connection of the BL with anchoring fibrils becomes visible (Fig. 80).

4 Discussion

4.1 The Nucleolar Channel System

With only one exception, the NCS has so far always been found in glandular epithelium of the human endometrium – it does not even occur in the endometrium of other primates (Mac-Lennan et al. 1971). The one exception reported so far is in the secretory cells of the human endocervix (Yasuzumi et al. 1981), although it does not occur in these cells until after discontinuance of a low-dose progestin (norethiosterone acetate). In the first paper on the NCS (Dubrauszky and Pohlmann 1960), it was pointed out that the NCS is formed during the secretory phase of the cycle, and since that time most authors have described the NCS as a postovulatory structure (Ancla and De Brux 1965; Kohorn et al. 1972; Gordon 1975; More et al. 1975; Spornitz 1987). Feldhaus et al. (1977), however, reported that the NCS is already present during the middle to late proliferative phase, but failed to indicate how the endometrium which they investigated was dated. When the endometrium is dated purely on the basis of biopsy material obtained through curettage, there are many inaccuracies (Honoré et al. 1988). There are always isolated tissue areas within the endometrium which do not respond properly to the circulating hormones. Moreover quantification of the progesterone level alone does not always suffice to determine the actual day of the cycle (Shepard and Senturia 1977). Even the distribution pattern of cytoplasmic receptors for progesterone and estrogen and their concentration varies greatly from area to area within the same endometrium (Tsibris et al. 1978). Therefore, the findings of Feldhaus et al. (1977) probably need reinterpretation. Recently it was shown that by determining the luteinizing hormone (LH) level the endometrium can be dated fairly accurately (Dockery et al. 1988a,b). Using this method these authors found that the NCS only occurred during the postovulatory phase of the cycle. In the present study it was clearly shown that the NCS was present in all of the investigated uteri between days 18 and 25. Maximum development of the NCS was reached between days 19 and 21, during which time the NCS, as a dynamic entity, consists of seven sets of three-branched tubules which are connected to the nuclear membrane by means of four contact points. The three rows of tubules run parallel to each other, without supercoiling. From the nuclear pole to the karyoplasmic pole, the spiral runs in one and a half turn. This model of the NCS is very different from the various models proposed by More et al. (1974) who, for their first model claimed that the NCS consisted of two tubules running parallel to each other, with each one ending blindly in the karyoplasm after eight turns.

Then 1 year later the same authors proposed that the NCS consisted of supercoiled tubules (More et al. 1975), only to replace this model in 1980 by a new one which supposedly had up to six individual tubules running across the core of the NCS. None of these proposed models was verified by either manual or computer-aided reconstruction. The structure proposed in 1980 by More et al. was based on 0.5-µm thick sections, and it might well be that the reported tubules running across the core, and the reported four to five rows of tubules running parallel to each other arise from superimposition caused by the thickness of the sections. The fully developed NCS (i.e., not regressing NCS) never consists of more than three rows of tubules, as previously described.

Manual reconstruction, as performed in my laboratory for the investigation of the NCS, is a rather time-consuming and painstaking procedure which is not applicable to the study of large numbers of objects. For this reason computer-aided reconstruction appears to be the method of choice, at least for partial reconstruction of such complex objects as the NCS. It is not only a relatively fast procedure but also lends itself to video animation, allowing the inspection of reconstructed objects from all possible angles of inclination.

The findings of Cornillie et al. that the NCS is not always associated with the nucleolus (1985) only holds true for the period between the 22nd day and desquamation. From the 1st day of its appearance up to day 21, the NCS is constantly associated with the nucleolus. It is only when regressive changes begin that the NCS may become detached from the nucleolus (see Fig. 11), but even then it is always bordered by RNP particles or ribosomes (Nakao et al. 1971; More et al. 1974, 1975).

To date, the physiological meaning or function of the NCS is not understood. The association of the NCS with the perinuclear space on the one hand and the nucleolus on the other indicates a possible connection with the synthesis and transport of RNA into the cytoplasm (More and McSeveny 1980). Direct evidence for this hypothesis, however, is still lacking. Ribosomal RNA has never been found within the lumen of the NCS tubules. Adenosine triphosphatase (ATPase) and 5′-nucleotidase have, however, been found within the NCS tubules (Buchwalow et al. 1985), which might also indicate the involvement of the NCS in RNA metabolism.

Invagination of the nuclear membrane at the site of NCS formation has been reported by a number of authors (Clyman 1963, Terzakis 1965, Sengel and Stoebner 1970). This invagination, which has been called a "cytoplasmic arm" (Terzakis 1965), is the area which gives the NCS the most direct access to the cytoplasm (Figs. 5, 9, 10), since after day 20 the NCS tubules are not always directly connected to the perinuclear space. Our present knowledge does not indicate that the NCS is in any way linked to the formation of giant mitochondria and subnuclear glycogen, since it is still in the process of formation at a time when mitochondrial transformation and glycogen synthesis are well under way (Verma 1983, Cornillie et al. 1985).

As previously mentioned the NCS was found in every investigated endometrium taken between days 18 and 25, with the exception of one. In this case the patient had reported anovulatory cycles long before hysterectomy – the cycle during which the uterus was taken might well have been anovulatory, too. The ovaries were not excised, so that it was not possible to either rule out or confirm the presence of a fresh corpus luteum. Because of the rather time consuming preparatory procedure necessary for TEM, it was only weeks after hysterectomy that the NCS was found to be missing when it was no longer sensible to establish hormonal profiles.

In contrast to the untreated uteri, the uteri of patients with progesterone IUDs never develop an NCS, not even in regions distant to the IUD, such as the areas of the tubal openings. The same is true for patients taking the "Minipill" (low-dose lynestrenol, Exluton, Organon, The Netherlands). Under this medication only 10%–20% of the cycles are reported to be anovulatory; in 80%–90% of the cycles ovulation is simply postponed until days 20–21 (Mall-Haefeli et al. 1976). One might, therefore, expect the NCS to develop after a delay of a few days; it was found, however, during the present study that the NCS does not form under the Minipill, not even towards the end of the cycle. In patients using sequential or step-up preparations which suppress ovulation, the NCS did not develop either.

In contrast, with patients using a copper IUD the NCS always developed, even in areas which are in direct contact with the copper wire. The released copper did, however, have an observable influence on the formation of the NCS inasmuch as the NCS was smaller than normal, and the number of rows of tubules was sometimes reduced to two.

In the case of aplastic ovaries, administration of estrogen and progesterone with subsequent IVF makes implantation of an embryo possible (Novat et al. 1986). Where this was achieved the NCS was always present, indicating that the NCS plays an important, if not central role, either in implantation itself or in the preparation of the endometrium for implantation. We hope that, in the near future, the NCS may be isolated by means of differential centrifugation, and its function elucidated by subsequent biochemical characterization.

4.2 Giant Mitochondria

The three-dimensional structure of mitochondria in human endometrium has so far been given very little attention. Changes in the three-dimensional structure of regular mitochondria during the course of a cycle were reported by Worth and Lucas (1978), who observed that the external form of mitochondria changes greatly during the cycle, ranging from spheric over tubular to complex mitochondria, the latter either having a U-, V-, or Y-like shape. Towards the end of the cycle the number of complex mitochondria increases. The giant mitochondria of the postovulatory triad – an important postovulatory characteristic of the glandular epithelium – are not even mentioned by the above authors.

Coaker et al. investigated complex giant mitochondria by serial sectioning and high voltage microscopy of 2 µm thick sections (Coaker et al. 1982). These authors believe giant mitochondria to be local swellings of exceptionally long and branched but otherwise normal mitochondria, and that during the phase of the cycle when giant mitochondria appear the total number of mitochondria is drastically reduced, sometimes to a point where only one singular complex mitochondrion is left per cell.

These findings in no way agreed with our own findings, and we therefore investigated a large number of mitochondria by means of computer-aided reconstruction. The results proved that there are many complex mitochondria present during the secretory phase of the cycle, which are not, however, part of a "singular mitochondrion." No such entity exists at any phase of the cycle. The observations reported by Coaker et al. can only be interpreted as resulting from the superimposition of mitochondria as it occurs in sections of such an enormous thickness as those used in that study.

The extensions of the complex mitochondria are probably given off in order to reduce mitochondrial size to normal. Towards the end of the cycle, giant mitochondria have practically vanished from the glandular epithelium.

4.3 Glycogen in Glandular Epithelial Cells

The postovulatory accumulation of glycogen in the subnuclear region of the glandular epithelial cells is an irrefutable fact. Little is known, however, about the "acceptor molecule" or "primer" of glycogenesis, without which glycogen synthetase is not able to form glycogen de novo (Ryman and Whelan 1971). With in vitro experiments a maximal rate of synthesis is achieved by the addition of glycogen as an acceptor molecule. Limit dextrins of glycogen are much less effective than native glycogen (Leloir 1964). Other molecules are also effective acceptors, including α-1,4-glucan protein (Krisman and Barengo 1975) and various oligosaccharides (Brown and Larner 1971). In muscle it has been shown that free glucose molecules can even attach to phosphorylase with subsequent further binding of glucose molecules. Finally an oligosaccharide molecule is detached from the enzyme which then acts as an acceptor molecule for long-chain synthesis of glycogen (Illingworth et al. 1961).

The participation of SER in glycogen metabolism has been discussed for many years (cf. review by Cardell 1977). The presence of glucose-6-phosphatase in the microsomal cell fraction indicates the participation of SER in glycogen breakdown rather than synthesis, a notion supported by the fact that glycogen synthetase is a cytosolic enzyme (Luck 1961). Thus, it was accepted for quite some years that the SER was involved solely in glycogen breakdown, and only later accepted that a morphological relationship exists between the ER and newly formed glycogen particles (Vrensen and Kuyper 1969; Cardell 1977; Spornitz 1978). Vrensen and Kuyper demonstrated that newly formed glycogen particles possess a high affinity to uranyl acetate, as is normally the case with ribosomes. They are also of the opinion that ribosomes are incorporated into newly formed glycogen, which subsequently detaches from the RER, thus forming SER. This would explain the close morphological relationship between ER membranes and newly formed glycogen particles which has been well documented in this paper. The relationship with ribosomal elements is particularly clear.

4.4 Decidualization

A large number of studies on decidualization have been done using rodents which have an estrous cycle, but not a real menstrual cycle. Thus, it is not surprising that these papers do not distinguish unambiguously between genuine decidual, predecidual and pseudodecidual cells (see Shelesnyak and Tic 1963; Finn 1971; Nilsson et al. 1978; O'Shea et al. 1983). The results of the present study clearly indicate that a strict differentiation between genuine decidual, predecidual and pseudodecidual cells should be made. Those cells formed towards the end of the menstrual cycle should be called predecidual cells, since they are the precursor cells of the genuine decidual cells of pregnancy. As described above, genuine decidual cells are characterized by four typical features: (a) a giant cell volume (the nuclear volume constituting approx. 6%–10% of the cell volume), (b) miniature mitochondria, (c) decidual granules, and (d) a continuous coat of BL. All these features must be present before the cells can be called genuine decidual cells, and, when these features are present, even cells from a pseudopregnancy or cells from an uterus under a progesterone IUD should be named genuine decidual cells. The decidual cells formed under a progesterone IUD differ only in terms of quantity and not in terms of quality. Thus, it seems appropriate to apply the term genuine decidual cell. In contrast to genuine decidual cells, all cells that do not possess the above four characteristic features, e.g., cells from tubal pregnancy (Spornitz 1991), should be named pseudodecidual cells.

The origin of fibrinoid has not yet been satisfactorily clarified. Since fibrinoid is particularly well developed in the area of the decidua basalis (Azab et al. 1972; Pijnenborg et al. 1980), it does, however, seem likely that the trophoblast plays a role in fibrinoid formation. The present study also revealed that fibrinoid was practically absent in the decidua parietalis. The decidua formed under a progesterone IUD is absolutely free from embryonal, i.e., trophoblastic influence, yet fibrinoid is still formed in great quantities. Thus, the trophoblast appears to help stimulate fibrinoid formation, rather than being a sine qua non. This hypothesis is supported by the fact that the amount of fibrinoid formed during pregnancy is much greater than that formed under a progesterone IUD.

The number of decidual granules on decidual cells from patients with a progesterone IUD is about four times as high as that on cells from pregnant patients. Since decidual granules are incorporated into the fibrinoid, as described in this study, it then follows that the presence or absence of trophoblast affects the secretion of decidual granules rather than having an effect on their synthesis. This would also explain the more pronounced formation of fibrinoid during pregnancy.

The progesterone IUD-induced decidua appears to be an ideal subject in the investigation of both fibrinoid and the maternal part of the uteroplacental interface. The thickness of the BL on decidual cells also reflects the presence or absence of trophoblast; it measures 420 nm on decidual cells from pregnancy compared with 100 nm on decidual cells from patients with a progesterone IUD. Cells from the decidua basalis, which also has very little influence from the trophoblast, also form a rather thin BL. The BL on decidual cells contains laminin, type IV collagen, fibronectin, and heparan sulfate proteoglycan (Wewer et al. 1985), which makes it comparable to the normal BL, found on epithelial cells (Inoué and Leblond 1988). As was demonstrated by Wewer et al. (1985), fibrinoid has a different composition, making the participatation of decidual granules in the formation of BL unlikely.

Collagen vacuoles as described in the present study were already described in decidual cells by Wessel in 1959. He believed these vacuoles were simple invaginations, but serial sectioning in this study has proved that they are intracellular structures. The appearance of collagen vacuoles – already present in predecidual cells – is accompanied by an increase in the amount of extracellular collagen. This suggests the participation of collagen vacuoles in the synthesis of collagen rather than in its degradation – at least initially. In later stages of pregnancy and after prolonged usage of a progesterone IUD, the number of collagen vacuoles decreases and lysosomal elements can be found inside the vacuoles together with apparently dissolving collagen fibrils. This is in good agreement with the reported postpartum degradation of uterine collagen in rats (Shimizu et al. 1983) and mice (Henell et al. 1983). Intracellular collagen has also been found in other tissues, mainly under experimental or pathological conditions. After ligature of ureteric arteries, the smooth muscle cells of the ureter developed collagen vacuoles which are comparable to the collagen vacuoles found in arteriosclerotic human vessels (Staubesand 1977). Fibroblasts of tendon and periodontium also possess collagen vacuoles (Chaplin and Greenlee 1975; Moon-Il and Garant 1981; Kopf and Staubesand 1984; Michna 1988). Staubesand (1977) described three possible functions of collagen vacuoles:

1. They promote the intracellular formation of collagen from tropocollagen molecules.
2. They reflect phagocytosis and intracellular degradation.
3. They reflect muscular degeneration.

The first two functions were still presented as the most probable explanation in a paper from the same laboratory published years later (Kopf and Staubesand 1984). Michna (1988) prefers to consider the presence of collagen vacuoles as degradative vacuoles which are formed to regulate the amount of collagen in the absence of other regulatory mechanisms. For the late decidua this might well be the case; predecidual cells, however, which are at the point of turning into decidual cells, still synthesize increasing amounts of collagen.

4.5 Endometrial Granulocytes (K Cells)

Granulated endometrial stroma cells were described properly for the first time by Weill (1921) who compared these cells to similar cells in the mucosal lining of the intestine. It is quite possible that the same cell might already have been described in earlier studies, but owing to the poor quality of the illustrations in some publications, the authors might just as well have described genuine leukocytes (Marchand 1904; Ulesco-Stroganoff 1908). The granules of the granulated endometrial stroma cells were later found to be phloxinophilic (Hamperl 1954; Hellweg 1954; Hamperl 1955; Hamperl and Hellweg 1958), and they apparently contain relaxin (RLX) (Dallenbach and Dallenbach-Hellweg 1964). This same hormone was also found in the granulated metrial gland cells of rats and mice (Dallenbach-Hellweg et al. 1965). The endometrium of primates other than man has also been found to contain a cell type which is very similar to the human granulated endometrial stroma cell (Cardell et al. 1969, Enders et al. 1985). Because of the granules they contain, these cells were named K cells (from the German word for granule, *Körnchen*).

The origin of K cells has always been of particular interest. Dallenbach-Hellweg believes that K cells originate from stromal fibroblast (1975), but since the transformation process has never been morphologically documented, this interpretation is not held in very high regard.

Various immunological studies have focused on the decidua in recent years, since the decidua is apparently involved in the suppression of immunological reactions between maternal and embryonal tissues (Kearns and Lala 1982). Placental grafts that were transplanted into uteri of ovariectomized mice penetrated into the myometrium; this, however, did not happen in decidualized uteri (Kirby and Cowell 1968, cited in Kearns and Lala 1982). While searching for immunocompetent cells which limit the invasion of embryonic cells into the maternal tissue, various investigators considered bone marrow cells (Kearns and Lala 1982; Peel et al. 1982; Bulmer and Sunderland 1983; Gambel et al. 1985).

Gambel et al. (1985), who believe that the immunocompetent cells of decidua are of bone marrow origin, do not differentiate between decidual cells and endometrial granulocytes. Kearns and Lala have already postulated that stromal fibrocytes, i.e., the stem cells of decidual cells, are derived from bone marrow. Bulmer and Sunderland (1983) believe that the presence of leukocyte antigen on human K cells is indicative of their bone marrow origin. When the bone marrow of lethally irradiated mice was substituted with rat bone marrow, rat-type endometrial granulocytes were subsequently found in deciduomata of these animals (Peel et al. 1982); this supports the hypothesis that "granulated metrial gland cells" might originate from lymphocytes (Peel and Bulmer 1977).

On the basis of morphological evidence documented in this book, the development of K cells from lymphocyte-like precursor cells appears to be very probable. The development from cells that are morphologically identical to lymphocytes to similar cells with K cell granules, and finally to fully mature K cells can be clearly observed in postovulatory endometrium and early decidua. In contrast, the postulated development of K cells from stromal fibrocytes (Dallenbach-Hellweg 1975) is a hypothesis totally unsupported by observations at any time during the menstrual cycle. The Dallenbach-Hellweg hypothesis is, in fact, based on light microscopic sections which are extremely difficult to interpret as far as subcellular changes are concerned.

Presupposing a lymphocytic origin, K cells might well be capable of regulating immunological mechanisms; they might aid the toleration of the implanting embryo and prevent the invasion of myometrium by the embryo. During weeks 8–18 of pregnancy, K cells are predominantly found in the region of the implanting trophoblast and immediately surrounding degenerating maternal tissue (Pijnenborg et al. 1980); this is reason enough for the authors to believe that these cells are involved in immunological mechanisms.

Active immunosuppression in the uteri of mice involves lymphocytes or lymphocyte-like cells. These cells do not carry T cells markers and they accumulate in the region of the uteroplacental interface (Clark et al. 1984).

As previously described, K cells are frequently found in close contact with predecidual cells. Associated with this contact is the formation of gap junctions, after which the predecidual cells develop into decidual cells. The gap junctions might serve as channels for the passage of some sort of "developmental or immunological information."

Towards the end of pregnancy the number of K cells is drastically reduced, mainly through their transformation into miniature cells and subsequent degradation. One would expect the RLX content of the decidua to be likewise reduced, as K cells are known to synthesize RLX; Bigazzi et al. (1980) indeed found that the RLX level of the decidua was at its lowest just before birth.

4.6 Prolactin from Decidua

Human amniotic fluid contains up to a hundred times more prolactin than is present in maternal blood (Tyson et al. 1972). Several studies have revealed that the decidua is the place where prolactin is synthesized (Riddick and Kusmsik 1977; Riddick and Daly 1982; Braverman et al. 1984; Hochner-Celnikier et al. 1984), and employing antiprolactin serum in immunohistochemistry, a distinct reaction was produced in decidual cells (Meuris et al. 1980). On the basis of chromatography, electrophoresis, immunoprecipitation, and receptor assay technic, decidua prolactin could not be distinguished from pituitary prolactin (Golander et al. 1978). Differential and density gradient centrifugation, however, showed that the subcellular compartment where prolactin is stored in decidual cells is different from that of the hypophysis (Handwerger et al. 1984), which might account for the different regulatory mechanisms which exist for both hormones (Handwerger et al. 1983). Pituitary prolactin is stored in membrane-bounded secretory granules (Kovacs et al. 1977), but decidual prolactin has not yet been localized on the subcellular level. Differential centrifugation indicates, however, that it might possibly be stored in a form other than secretory granules, as it sediments with the post-microsomal fraction of the cell (Handwerger et al. 1984). Prolactin also occurs in normal endometrium towards the end of the secretory phase, i.e., when predecidual cells have been formed (Maslar and Riddick 1979). The amount of prolactin synthesized seems to be dependent upon the degree of decidualization and upon the level of circulating progesterone (Daly et al. 1983). Decidual granules can probably be ruled out as storage sites for prolactin as it is already present in high amounts in predecidual cells (Maslar and Riddick 1979). Moreover, prolactin can be synthesized in cells which are free of decidual granules (Hochner-Celnikier et al. 1984).

The secretory granules of decidual cells, which are very different from decidual granules, as previously described, might well be related to decidual prolactin. The contents of these granules can be digested with proteolytic enzymes and identical granules are, in fact, found in late predecidual cells (Spornitz unpublished observation). In an immunohistochemical study of prolactin it was found that it is unevenly distributed in the late secretory endometrium (McRae et al. 1986). This is in good agreement with our own findings that secretory granules are unevenly distributed within the predecidual cells of the late secretory phase.

4.7 Desquamation

At first glance, long-term effects of IUDs, as described in this study, appear to be quite understandable. When the dynamics of the endometrium, including desquamation and reepithelization, is, however, taken into consideration, the idea that there should be any long-term effects becomes rather irritating. Oral contraceptives act on the hypothalamohypophysial axis (Lyon et al. 1976; Kreutner et al. 1976; Wittlinger 1980; Ludwig 1982; Mall-Haefeli 1986) as do contraceptive implants (Keller 1986). Consequently, these systemic contraceptives also act on the ovaries, affecting the levels of estrogen and progesterone. Any long-term effects on the endometrium are, therefore, understandable.

In a study on long-term effects of inert IUDs (Lippes loop), two long-term effects were observed: inflammatory reactions and asynchronous development of the endometrium (Lee et al. 1983). The inflammatory reaction might well be caused by a secondary infection which is not due to contact with the IUD but to ascending vaginal bacteria which are reported to reach the uterine cavity by means of the IUD tail (Spornitz et al. 1984). Since inert IUDs do not release any bioactive substances, as do medicated IUDs, the asynchronous development of the endometrium which occurs with long-term usage of IUDs is hard to explain.

The progesterone IUD acts through the local release of progesterone (Pharriss et al. 1974). A systemic action of the locally released hormone is, however, considered to be possible. The ultrastructural morphology of fallopian tubes in patients with progesterone IUDs is reported to be similar to that of patients during the first trimester of pregnancy (Spornitz et al. 1980; Mall-Haefeli and Spornitz 1984). During lactation it was found that the number of anovulatory cycles amounts to 100% among progesterone IUD users, as opposed to only 33% among copper IUD users (Abdalla et al. 1981). The hormonal profiles of progesterone IUD users change after several months (Mall-Haefeli et al. 1977; Mall-Haefeli 1984), although it appears doubtful that hormonal changes are the only cause of the long-term effects previously described (see Sect. 3.3).

One possible mechanism which would explain these long-term effects could be an incomplete desquamation; this would expose the remaining areas of endometrium to the prolonged influence of mechanical trauma in the case of inert IUDs, or of bioactive substances in the case of medicated IUDs. The true extent of normal desquamation has been discussed frequently (cf. Bartelmez 1957). This author doubted that the endometrium is fully desquamated during menstruation and believed that renewal of the surface epithelium is achieved through the outgrowth of remaining (not desquamated) surface epithelium. Other authors believe that the stroma is the source of re-epithelialization (Baggish et al. 1967). Ferenczy is of the opinion that the remaining glandular stumps and the region of the tubal openings is the main source of re-epithelialization (Ferenczy 1976a,b; Ferenczy et al. 1979a). The rate of mitosis was not, however, substantially elevated, which led Ferenczy to the conclusion that re-epithelialization is achieved through endomitotic cell division (Ferenczy et al. 1979b). Ludwig and Metzger (1976b) found that re-epithelialization started from the remaining parts of the endometrial glands, and Wagner and Beller (1979) observed that not only glandular stumps but also remaining surface epithelium and intact glands participated in endometrial regeneration. These authors believe that the extent of desquamation has become a question of semantics. The above-mentioned papers do, however, indicate that desquamation generally is not complete, leaving varying areas of surface and glandular epithelium intact. In most cases re-epithelialization has started already before desquamation has been finished (Ludwig and Metzger 1976b; Wagner and Beller 1979). Thus, it is extremely difficult to quantify regeneration and desquamation. As Flowers and Wilborn (1979) stated: "at no time during menstruation was there complete denudation of the surface epithelium."

With the progesterone IUD desquamation seems to be influenced by yet another factor: large areas of the endometrium are decidualized and are thus exempted from cyclic changes. This might be a possible explanation for the observed reduction in the amount of menstrual blood flow by 41% as compared with copper IUD users (Rybo and Bergqvist 1976).

The decidualized areas are, therefore, exposed to the released progesterone for many months, if not years, which might account for the observed glandular atrophy, exhaustion of decidual cells, and degradation of K cells as described in this paper.

4.8 The Endometrium Under IUDs

Up to the present day the mode of action of IUDs is not fully understood (cf. Johannisson 1987; Toder et al. 1988). Various mechanism have been discussed, most of which are based on the suppression of implantation (Moyer and Shaw 1984; Elstein and Ezenyirioha 1984).

As far as our own results are concerned, decidualization appears to be the number one factor in the contraceptive action of progesterone IUDs. A similar decidualization has been reported with other progestins like the synthetic D-norgestrel (Nilsson et al. 1978) or levonorgestrel (Wadsworth 1980).

With copper IUDs, the FBR resulting in a massive infiltration by leukocytes appears to be the major contraceptive factor. In mice there is a reciprocal relationship between contraceptive efficacy of IUDs and the number of leukocytes present within the uterine cavity (Toder 1988). Another important factor seems to be the desynchronization between blastocyst development and maturation of the endometrium (see Sect. 3.4).

Particular attention should be paid to the formation of micropolyps under copper IUDs, which so far has only been noticed by a small number of authors (Ludwig 1980; De Brux et al. 1981; Ludwig and Metzger 1984; Spornitz 1985). These micropolyps are most likely a reaction of the endometrium to the mechanical trauma of the IUD; they are occasionally also formed under progesterone IUDs (Spornitz, unpublished observation). Since the micropolyps are already present during the early proliferative phase, immediately after desquamation, it is possible that they have escaped desquamation.

Intermenstrual bleeding or spotting is among the most frequent side effects of IUDs (Tauber and Nohlen 1984). The micropolyps which are vascularized are particularly well developed in the areas of direct contact with the IUDs. This is also the tissue zone of maximal mechanical stimulus, evidenced by the presence of imprints in this area, and it might well be that torn off or injured micropolyps are the source of intermenstrual bleeding. Increased fibrinolytic activity as observed under copper IUDs (Larsson et al. 1975; Liedholm et al. 1983) certainly also plays a role in intermenstrual and irregular menstrual bleedings (Johannisson 1987).

4.9 Conclusions

The postovulatory triad, comprising subnuclear glycogen, giant mitochondria, and the NCS, is fully developed before eutopic implantation takes place. The NCS seems to play an important role in the events leading to implantation, since it fails to develop under low-dose gestagen therapy (Minipill) with slightly postponed ovulation, under a progesterone IUD, or in anovulatory cycles. Giant mitochondria and subnuclear glycogen are probably just accompanying phenomena, rather than significant factors in implantation. By isolating the NCS with differential or density gradient centrifugation and identifying it biochemically, we hope that we may be able, in the near future, to better understand the exact role played by the NCS. Should a central role in implantation be substantiated, selective suppression of the NCS could be used to block implantation.

If the NCS is really a sine qua non for eutopic implantation, the question remains as to how ectopic implantion is possible, when there are no K cells, no NCS, and no predecidual cells? Could it be that the NCS, K cells, and predecidual cells are necessary for the maintenance of pregnancy and for the regulation of immunological mechanisms rather than being perequisites for implantation? If that is the case, then surface phenomena are probably responsible for the attachment of the blastocyst in eutopic as well as ectopic pregnancy. This would certainly necessitate a reinterpretation of the various mechanisms discussed for contraceptives.

5 Summary

The functional morphology of the human endometrium and decidua has been investigated with particular attention to all aspects of implantation and contraception. The postovulatory triad (of glandular epithelium), comprising subnuclear glycogen, giant mitochondria, and the NCS seems to be involved in both implantation and contraception and has been studied in detail.

Initial stages of glycogen formation were found to be related to straight, uncoiled forms of polysomes which, by acting as acceptor molecules for glucose start glycogen synthesis. Discharge of glycogen from the glandular epithelium is achieved by a type of apocrine secretion.

Computer-aided reconstruction of giant mitochondria, based on ultra-thin serial sections, revealed that giant mitochondria of the secretory phase of the cycle are not comprised of a single complex mitochondrion, but of numerous single mitochondria. Towards the end of the cycle, these mitochondria either reduce their size to normal by means of budding of protrusions or they are degraded in autophagic vacuoles.

The NCS was studied for the first time from the initial stages of formation to the final stages of intracellular degradation. Based on ultra-thin serial sections the NCS was reconstructed with manual and computer-aided reconstruction. It was found to consist of seven sets of tubules, each containing three rows of coiled tubules which are wound in one-and-a-half turns from the site of origin at the nuclear membrane to the karyoplasmic pole. The individual tubules are interconnected by means of lacunae or connecting tubules. The formation of the NCS is suppressed by progesterone IUDs, but not by copper IUDs, and it is not formed in anovulatory cycles or under low-dose gestagen therapy (Minipill). The NCS is believed to participate in regulatory mechanisms of the glandular epithelium. Thus, the contraceptive action of the progesteronelUD may in part be due to the suppression of the NCS and not only due to decidualization of the endometrial stroma.

From the middle to late secretory phase K cells (endometrial granulocytes) and predecidual cells are formed. Their development and ultimate fate was studied during the normal physiological cycle, under a progesterone IUD, and also during pregnancy. K cells do not develop from stromal fibrocytes, as has been postulated in the past, but from lymphocyte-like cells which are present in the endometrium after ovulation. These cells multiply through mitotic division within the endometrial stroma. Under a progesterone IUD and during pregnancy they discharge their secretory granules and reduce their size drastically by giving off glycogen-containing cytoplasmic patches. During pregnancy they reduce their size by simultaneously fractionating their nuclei, thereby forming miniature cells with a total diameter of 1.6–2.4 µm. Towards the end of pregnancy they disappear totally. Decidual cells were studied in detail and were found to possess four typical characteristics which clearly distinguish them from other cells in the decidua:

1. A giant cell volume (the nuclear volume constituting approx. 6%–10% of the cell volume).
2. Miniature mitochondria.
3. Decidual granules.
4. A continuous BL.

Decidual cells from the decidua basalis, parietalis, and capsularis differ greatly in shape and structure, as demonstrated in morphometric studies. The amount and composition of intercellular substance also differs greatly within the three decidual regions. A new kind of secretory granule which is not related to decidual granules was found in decidual cells. These secretory granules are morphologically similar to hypophysial prolactin granules, and are digestible with proteolytic enzymes. Their possible relation to decidual prolactin is discussed.

The effect of copper and progesterone IUDs on the morphology of the endometrium was studied with both SEM and TEM. It was found that copper IUDs induce the formation of vascularized micropolyps which might be responsible for intermenstrual bleeding and spotting. The mode of action of both IUDs is discussed. The decidualized cells of tubal pregnancy are of a columnar cytotrophoblastic origin and do not possess any of the typical features of genuine decidual cells. It is, therefore, suggested that these cells should be called pseudodecidual cells.

Acknowledgements. I am indebted to Prof. Dr. K.S. Ludwig who first instilled in me a fascination with the human endometrium, and am grateful for the many years of stimulating interest he showed in my work.

I am thankful to Prof. Dr. D. Sasse for his understanding and encouraging criticism of my work, and to Prof. Dr. Annetrudi Kress for many a fruitful discussion.

Thanks go Prof. Dr. M. Mall-Haefeli, Prof. Dr. H. Ludwig, Dr. F. Allemann, Prof. Dr. H.S.E. Hafez, Dr. F. Lasagni, Dr. U. Lüscher, Dr. O. Cimber for endometrial biopsies and hysterectomized uteri. The much-appreciated help of Prof. Dr. O. Kübler and Dr. J. Ylä-Jääski of the "Institut für Kommunikationstechniken, Fachgruppe für Bildwissenschaften, ETHZ", in reconstructing the NCS and the mitochondria with the aid of a computer is gratefully acknowledged.

My special thanks go to the staff, past and present, of the Department of Electron Microscopy, Institute of Anatomy, Basel. In particular I wish to thank Gianni Morson for his excellent and never-tiring help with all technical aspects of electron microscopy, from maintenance of equipment to manual reconstruction of the NCS. My thanks goes also to Mrs. I. Bartuskova for her skillful preparation of SEM specimen, to Mrs. H. Schaller for histological preparation, and to Mr. H. J. Stöcklin for his excellent photographic work.

References

Abdalla MI, Ibrahim II, Osman MI, Badraoui MHH, Askalani H (1981) Corpus luteum function in lactating women using the progestasert system. Contracept Deliv Syst 2:127–132

Ancla M, De Brux J (1965) Occurrence of intranuclear tubular structure in human endometrium during the secretory phase, and of annulate lamellae in hyperestrogenic states. Obstet Gynecol 26:23–33

Ancla M, de Brux J, Belaisch J, Musset R (1964) Influence de l'equilibre oestrogene-progesterone sur les ultrastructures de l'endometre humain. Gyn Obst 63:239–248

Armstrong EM, More IAR, McSeveney D, Chatfield WR (1973) Reappraisal of the ultrastructure of the human endometrial glandular cell. J Obstet Gynaecol Brit Comm 80:446–460

Asplund J, Holmgren H (1947) On the occurrence of metachromatically granulated cells in the mucous membrane of the human uterus. Acta Anat 3:312–330

Astwood EB (1938) A six-hour assay for the quantitative determination of estrogen. Endocrinology 23:25–31

Azab I, Okamura H, Beer A (1972) Decidual cell production of human placental fibrinoid. Obstet Gynecol 40:186–193.

Baggish MS, Pauerstein CJ, Woodruff JD (1967) Role of stroma in regeneration of endometrial epithelium. Am J Obstet Gynecol 99:459–465

Bartelmez GW (1957) The phases of the menstrual cycle and their interpretation in terms of the pregnancy cycle. Am J Obstet Gynecol 74:931–955

Beller FK, Schweppe KW, Wagner H (eds) (1984) Intrauterinpessare. Edition Medizin Weinheim

Bernhard W (1968) Une methode de coloration regressive a l'usage de la microscopie electronique. C.R. Acad Sci Ser D 267:2170–2173

Berthillier G, Morelis P, Colobert L, Trefouel J (1969) Mise en evidence de glycogene lie aus membranes des vesicules microsomiques de foie de rat. C.R. Acad sci (Paris) 268:1793–1794

Bigazzi M, Nardi E, Bruni P, Petrucci F (1980) Relaxin in human decidua. J Clin Endocrin Metab 51:939–941

Blandau RJ (1961) Biology of eggs and implantation. In: Young WC (ed) Sex and internal secretions, 3rd edn: vol 2. Williams and Wilkins, Baltimore, pp 797–882

Braverman MB, Bagni A, Ziegler D, Den T, Gurpide E (1984) Isolation of prolactin-producing cells from first and second trimester decidua. J Clin Endocrin Metabol 58:521–525

Brown NE, Larner J (1971) Molecular characteristics of the totally dependent and independent forms of glycogen synthase of rabbit skeletal muscle. I. Preparation and characteristics of the totally glucose 6-phosphate dependent form. Biochim Biophys Acta 242:69–80

Buchwalow IG, Belyaeva LA, Zavalishina, LE (1985) Localization of nucleoside phosphatases (ATPase and 5′-nucleotidase) and nuclear ribonucleoproteins in the human endometric glandular cells during the secretory phase. Acta Histochem 77:205–208

Bulmer JN, Sunderland CA (1983) Bone marrow origin of endometrial granulocytes in the early human placental bed. J Reprod Immunol 5:383–387

Cardell RR (1971) Action of metabolic hormones on the fine structure of rat liver cells. I. Effect of fasting on the ultrastructure of the hepatocytes. Am J Anat 131:21–54

Cardell RR (1977) Smooth endoplasmic recticulum in rat hepatocytes during glycogen deposition and depletion. Int Rev Cytol 48:221–279

Cardell RR Jr, Hisaw FL, Dawson AB (1969) The fine structure of granular cells in the uterine endometrium of the rhesus monkey (Macaca mulatta) with a discussion of the possible function of these cells in relaxin secretion. Am J Anat 124:307–340

Cavazos F, Lucas FV (1973) Ultrastructure of the endometrium. In: Norris HJ, Hertig AT, Abell MR (eds) The uterus. Williams and Wilkins, Baltimore 136–174

Cavazos F, Green JA, Hall DG, Lucas, FV (1967) Ultrastructure of the human endometrial glandular cell during the menstrual cycle. Am J Obstet Gynecol 99:833–854

Chaplin DM, Greenlee TK (1975) The development of human digital tendons. J Anat 120:253–274

Clark DA, Slapsys R, Croy BA, Krceck J, Rossant J (1984) Local active suppression by suppressor cells in the decidua. Am J Reprod Immunol 5:78–83

Clark JH, Hsueh AJW, Peck EJ (1977) Regulation of estrogen receptor replenishment by progesterone. Ann NY Acad Sci 286:161–179

Clyman, MJ (1963) A new structure observed in the nucleolus of the human endometrial epithelial cell. Am J Obstet Gynecol 86:430–432

Coaker T, Downie T, More IAR (1982) Complex giant mitochondria in the human endometrial glandular cell: serial sectioning, high voltage electron microscopic, and three dimensional reconstruction studies. J Ultrastruct Res 78:283–291

Cohen AL (1979) Critical point drying: principles and procedures. SEM II: 303–324

Cordier R, Snoeck J (1949) Deziduale Reaktion des Stroma ovarii waehrend der Schwangerschaft. Gynaecologia 128:17–23

Cornillie FJ, Lauweryns JM, Brosens IA (1985) Normal human endometrium. Gynecol Obstet Invest 20:113–129

Dallenbach FD, Dallenbach-Hellweg G (1964) Immunhistologische Untersuchungen zur Lokalisation des Relaxins in menschlicher Placenta und Decidua. Virch Arch Path Anat 337:301–316

Dallenbach-Hellweg G (1975) Histopathology of the endometrium. Springer, Berlin Heidelberg New York

Dallenbach-Hellweg G (1984) The endometrium of infertility. A review. Path Res Pract 178:527–537

Dallenbach-Hellweg G, Battista JV, Dallenbach FD (1965) Immunohistological and histochemical localisation of relaxin in the metrial gland of the pregnant rat. Am J Anat 117:433–450

Dallenbach-Hellweg G, Hohagen F, Hohlweg-Majert P (1984) Die histologische Reaktion des Endometriums auf Sondenreizung im Hinblick auf die Problematik des Embryonentransfers bei der In-vitro-Fertilisation. Geburtsh Frauenheilk 44:381–387

Daly DC, Maslar IA, Riddick DH (1983) Prolactin production during in vitro decidualization or proliferative endometrium. Am J Obstet Gynecol 145:672–678

De Brux J, Trevoux R, Chambe G (1981) Histological responses of the endometrium to the IUD. Contracept Deliv Syst 2:145–156

De Feo VJ (1963) Determination of the sensitive period for the induction of deciduomata in the rat by different inducing procedures. Endocrinology 73:488–497

De Feo VJ (1967) Decidualization. In: Wynn RM (ed) Cellular biology of the uterus. Appleton-Century-Crofts, pp 191–290

Dmowski WP, Rao R, Scommenga A (1980) The luteinized unruptured follicle syndrome and endometriosis. Fertil Steril 33:30–34

Dockery P, Li TC, Rogers AW, Cooke ID, Lenton EA (1988a) The ultrastructure of the glandular epithelium in the timed endometrial biopsy. Human Reproduction 3:826–834

Dockery P, Li TC, Rogers AW, Cooke LD, Lenton EA, Warren MA (1988b) An examination of the variation in timed endometrial biopsies. Human Reproduction 3:715–720

Dougherty CM, Anderson MR (1964) Endometriosis and andenomyosis. Am J Obstet Gynecol 89:23–40

Dubrauszky V, Pohlmann G (1960) Veraenderungen der submikroskopischen Struktur von Druesenzellen der Corpusmucosa waehrend des Zyklus. Proc Europ Reg Conf Electr Micr (Delft) II, pp 862–865

Dubrauszky V, Pohlmann G (1961) Die Ultrastruktur des Korpusendometriums waehrend des Cyklus. Arch Gynaek 196:180–199
Edwards RG, Fishel SB (1982) The human uterus in luteal phase and early pregnancy. In: Edwards RG, Purdy JM (eds) Human conception in vitro. Academic Press, London, pp 257–288
Edwards RG, Purdy JM (eds) (1982) Human conception in vitro. Academic Press, London
El-Badrawi HH, Hafez ESE (1980) Physiological mechanisms of IUDs. In: Hafez ESE, Van Os WAA (eds) Medicated intrauterine devices. Martinus Nijhoff, Den Haag, pp 60–83
El-Badrawi HH, Barnhart MI, Hafez ESE, Badraoui MHH, Fayad M, Shafeek A (1980) Effect of medicated IUDs on the endometrium as judged by scanning and transmission electron microscopy. Contracept Deliv Syst 1:212–213
Elstein M, Ezenyirioha M (1984) Hypothesen zur kontrazeptiven Wirkung von Intrauterinpessaren. In: Beller FK, Schweppe KW, Wagner H (eds) Intrauterinpessare. Edition Medizin Weinheim, pp 37–48
Enders AC, Welsh AO, Schlafke S (1985) Implantation in the rhesus monkey: endometrial responses. Am J Anat 173:147–169
Feldhaus FJ, Themann H, Wagner H, Verhagen A (1977) Feinstrukturelle Untersuchungen ueber das Nuclear-Channel-System im menschlichen Endometrium Arch Gynaekol 223:195–204
Ferenczy A (1976a) Studies on the cytodynamics of human endometrial regeneration. I. Scanning electron microscopy. Am J Obstet Gynecol 124:64–74
Ferenczy A (1976b) Studies on the cytodynamics of human endometrial regeneration. II. Transmission electron microscopy. Am J Obstet Gynecol 124:582–595
Ferenczy A, Richart RM (1974) Female reproductive system: dynamics of scan and transmission electron microscopy. John Wiley, New York
Ferenczy A, Bertrand G, Gelfand MM (1979a) Studies on cytodynamics of human endometrial regeneration. III. In vitro short term incubation historadiography. Am J Obstet Gynecol 134:297–304
Ferenczy A, Bertrand G, Gelfand MM (1979b) Proliferation kinetics of human endometrium during the normal menstrual cycle. Am J Obstet Gynecol 133:859–867
Feyrter F (1957) Ueber den zelligen Bestand des Stroma der menschlichen Corpusmucosa. Arch Gynaek 190:47–82
Finn CA (1971) The biology of decidual cells. In: Bishop, MWH. (ed) Advances in Reproductive Physiology, vol 5. Logas, London, pp 1–26
Finn CA, Keen PM (1962) Studies on deciduomata formation in the rat. J Reprod Fertil 4:215–216
Finn CA, Keen PM (1963) The induction of deciduomata in the rat. J Embryol Exp Morph 11:673–682
Flickinger GL, Elsner C, Illingworth DV, Muechler EK, Mikhail G (1977) Estrogen and progesterone receptors in the female genital tract of humans and monkeys. Ann NY Acad Sci 286:180–188
Flowers CE, Wilborn WH (1979) Histoenzymology of the human endometrium. In: Beller FK, Schumacher GFB (eds) The biology of the fluids of the female genital tract. Elsevier/North Holland, New York, pp 203–223
Fraenkel L (1903) Die Function des Corpus luteum. Arch Gynaekol 68:438–545
Gambel P, Rossant J, Hunziker RD (1985) Origin of decidual cells in murine pregnancy and pseudopregnancy. Transplantation 39:443–444
Gebhard C (1899) Pathologische Anatomie der weiblichen Sexualorgane. S Hirzel, Leipzig
Golander A, Hurley T, Barret J, Hizi A, Handwerger S (1978) Prolactin synthesis by human chorion-decidual tissue: a possible source of prolactin in the amniotic fluid. Science 202:311–312
Gompel C (1964) Structure fine des mitochondries de la cellule glandulaire endometriale humaine au cours du cycle menstruel. J Microscopie 3:427–436
Gordon M (1975) Cyclic changes in the fine structure of the epithelial cells of human endometrium. Int Rev Cytol 42:127–172
Gordon M, Kohorn EI, Gore BZ, Rice SI (1973) Effect of postovulatory oestrogens on the fine structure of the epithelial cell in human endometrium. J Repord Fert 34:375–378

Gould SF, Shannon JM, Cunha GR (1983) Nuclear estrogen binding sites in human endometriosis. Fertil Steril 39:520–524

Gregoire AT, Ledger WD, Moran MJ (1973) The glycogen content of the human female genital tract in cycling, menopausal, and women with endometrial and cervical carcinoma. Fertil Steril 24:198–201

Hackl H (1971) Zur homronellen Steuerung der Glykogensynthese im humanen Endometrium. Arch Gynaek 211:617–626

Hamilton WJ, Boyd JD, Mossman HW (1976) Human Embryology. MacMillan, London

Hamperl H (1954) Ueber endometriale Granulocyten (Endometriale Koernchenzellen). Klin Wochenschr 32:665–668

Hamperl H (1955) The granular endometrial stroma cells: a new cell type. J Path 69:358–359

Hamperl H, Hellweg G (1958) Granular endometrial stroma cells. Obstet Gynecol 11:379–387

Handwerger S, Barry S, Markoff E, Barret J, Conn PM (1983) Stimulation of the synthesis and release of decidual prolactin by a placental polypeptide. Endocrinology 112:1370–1374

Handwerger S, Wilson S, Conn MP (1984) Different subcellular storage sites for decidual- and pituitary derived prolactin: possible explanation for differences in regulation. Mol Cell Endocrinol 37:83–87

Hayat MA (1981) Fixation for electron microscopy. Academic, New York

Hefnaw F, Segal SJ (eds) (1975) Analysis of intrauterine contraception. Proceedings of the third international conference on intrauterine contraception, Cairo, 12–14 Dec 1974. North Holland, Amsterdam

Hellweg G (1954) Ueber endometriale Koernchenzellen (endometriale Granulocyten). Arch Gynaekol 185:150–166

Hellweg G (1959) Ueber koernchenhaltige Zellen im menschlichen und tierischen Endometrium (Endometriale Koernchenzellen, metachromasierende Zellen). Z Zellforsch 49:555–568

Henell F, Ericsson JLE, Glaumann H (1983) An electron microscopic study of the post-partum involution of rat uterus. Virchows Arch [B] 42:271–287

Herr JC, Heidger PM Jr, Scott JR, Anderson JW, Curet LB, Mossman HW (1978) Decidual cells in the human ovary at term: I. Incidence, gross anatomy and ultrastructural features of merocrine secretion. Am J Anat 152:7–28

Herr JC, Fares NH, Banhawy MA (1987) Decidual cells in the human ovary at term: II. Morphometric analysis of cytoplasmic processes and organelles. Am J Anat 179:277–284

Hitschmann F, Adler L (1908) Der Bau der Uterusschleimhaut des geschlechtsreifen Weibes mit besonderer Beruecksichtigung der Menstruation. Monatsschr Geburtsh Gynaekol 27:1–82

Hochner-Celnikier D, Ron M, Eldor A, Segal S, Palti Z, Fuks Z, Vlodavsky I (1984) Growth characteristics of human first trimester decidual cells cultured in serum free medium: production of prolactin prostaglandins and fibronectin. Biol Reprod 31:827–836

Hoeglund A, Odeblad E (1977) Sperm pentration in cervical mucus, a biophysical and group-theoretical approach. In: Insler V, Bettendort G (eds) The uterine cervix in reproduction. Thieme Stuttgart, pp 129–134

Honore LH, Cumming DC, Fahmy N (1988) Significant difference in the frequency of out-of-phase endometrial biopsies depending on the use of the Novak curette or the flexible polypropylene endometrial biopsy cannula (Pipelle). Gynecol Obstet Invest 26:338–340

Hughes EC, Jacobs RD, Rubulis A, Husney RM (1963) Carbohydrate pathways of the endometrium. Effects of ovular growth. Am J Obstet Gynecol 85:594–609

Hughes EC, Demers LM, Csermely T, Jones DB (1969) Organ culture of human endometrium. Effect of ovarian steroids. Am J Obstet Gynecol 105:707–720

Illingwworth B, Brown DH, Cori CF (1961) The de novo synthesis of polysaccharide by phosphorylase. Biochemistry 47:469–478

Inoue S, Leblond CP (1988) Three-dimensional network of cords: the main component of basement membranes. Am J Anat 181:341–358

Jelinek J, Ylikorkala O, Jelinekova M, Jarvinen PA, Alapiessa U (1978) Effect of endogenous and exogenous progesterone on human endometrial enzymes. Int J Fertil 23:31–37
Johannisson E (1987) Mechanism of action of intrauterine devices: biochemical changes. Contraception 36:11–22
Jones HW (1982) The ethics of in-vitro fertilization. In: Edwards RG, Purdy JM (eds) Human Conception in vitro. Academic, London, pp 351–357
Kearns M, Lala PK (1982) Bone marrow origin of decidual cell precursors in the pseudopregnant mouse uterus. J Exp Med 155:1537–1554
Keller PJ (1986) Hormonale Kontrazeption mit Depot-Praeparaten. Ther Umsch 43:395–399
Kindermann G (1988) Endometriose. Wesen und Entstehung. In: Friedberg V, Thomsen K (eds) Gynaekologie und Geburtshilfe. Thieme Stuttgart, pp 13.0–13.11. (Spezielle Gynaekologie 2, vol III/2)
Kinoshita S (1898) Ueber grosszellige deciduazellenaehnliche Wucherungen auf dem Peritoneum und den Ovarien dei intrauteriner Schwangerschaft. Monatsschr Geburtsh Gynaek 8:500–509
Klein F, Kuebler O (1986) Fast direct display of discrete volume data. Proc 8th Int Conf Pattern Recognition, Paris, pp 633–635
Kohorn EI, Rice SI, Hemperly S, Gordon M (1972) The relation of the structure of progestational steroids to nuceolar differentiation in human endometrium. J Clin Endocr 34:257–264
Kopf M, Staubesand J (1984) Intrazelluläres Kollagen in Fibrocyten der Achillessehne junger Ratten. Acta Anat 120:39
Kovacs K, Horvath E, Higgins HP, Hudson AR, Gezina I, Ryan N (1977) Pituitary prolactin cell adenoma. Case histories in human medicine 15. Philips, Eindhoven
Kreutner A, Johnson D, Williamson HO (1976) Histology of the endometrium in long-term use of a sequential oral contraceptive. Fertil Steril 27:905–910
Krisman CR, Barengo R (1975) A precursor of glycogen biosynthesis: alpha-1, 4-glucan-protein. Eur J Biochem 52:117–123
Kuzela DC, Speers WD (1985) Heterotopic endometrium of the fallopian tube. Fertil Steril 44:552–553
Lanzavecchia G, Morano E (1958) Ultrastruttura della cellula ghiandolare dell'endometrio umano. Minerva Ginecol 10:882–884
Larkin LH, Cardell RR Jr (1971) Differentiation of granulated metrial cells in the uterus of the pregnant rat: an electron microscopic study. Am J Anat 132:241–258
Larsson B, Liedholm P, Astedt B (1975) Increased fibrinolytic activity in the endometrium of patients using copper-IUD (Gravigard). Int J Fertil 20:77–80
Lawn AM, Wilson EW, Finn CA (1971) The ultrastructure of the human decidual and predecidual cells. J Reprod Fert 26:85–90
Leavitt WW, Chen TJ, Allen T (1977) Regulation of progesterone receptor formation by estrogen action. Ann NY Acad Sci 286:210–225
Lee YB, Kim CS, Ro JY, Kwak HM, Song CH (1983) Long term effects of IUD on the human endometrium. Yonsei Med J 24:141–148
Leloir LF (1964) Role of uridine diphosphate glucose in the synthesis of glycogen. In: Whelan WJ, Cameron MP (eds) Control of glycogen metabolism. Churchill, London, pp 68–86 (Ciba Foundation Symposium)
Liebig W, Stegner HE (1977) Die Dezidualisation der endometrialen Stromazelle. Arch Gynaek 223:19–31
Liedholm P, Srivasta K, Wingerup L, Astedt BJ (1983) Higher fibrinolytic activity in human endometrium in direct contract with an IUD. Acta Obstet Gynecol Scand 62:169–170
Loeb L (1908) The production of deciduomata and the relation between the ovaries and the formation of the decidua. JAMA 50:1897–1901
Luck DJL (1961) Glycogen synthesis from uridine diphosphate glucose. The distribution of the enzyme in liver cell fractions. J Cell Biol 10:195–209
Ludwig H (1980) Intrauterinpessare: Veraenderungen am Endometerium und am Pessar nach verschieden langer Liegezeit in situ. Verh Dtsch Ges Path 64:523–535

Ludwig H (1982) The morphologic response of the human endometrium to long-term treatment with progestational agents. Am J Obstet Gynecol 142:796–808
Ludwig H, Metzger H (1976a) The human female reproductive tract. A scanning electron microscopic atlas. Springer, Berlin Heidelberg New York
Ludwig H, Metzger H (1976b) The re-epithelization of endometrium after menstrual desquamation. Arch Gynaek 221:51–60
Ludwig H, Metzger H (1984) Morphologische Aspekte der intrauterinen Kontrazeption. In: Beller FK, Schweppe KW, Wagner H (eds) Intrauterinpessare. Edition Medizin Weinheim, pp 12–22
Ludwig H, Metzger H, Hafez ESE (1976) Critical point drying and gold sputtering as applied to scanning electron microscopy of human reproductive tissues. Acta Anat 96:469–477
Ludwig KS, Spornitz UM (1977) Veraenderungen des Endometriums beim Progestasert-System. Acta Anat 99:237
Lyon FA, Frisch MJ (1976) Endometrial abnormalities occurring in young women on long-term sequential oral contraception. Obst Gynecol 6:639–643
MacLennan AH, Harris JA, Wynn RM (1971) Menstrual cycle of the Baboon: II. Endometrial ultrastructure. Obstet Gynecol 38:359–374
Mall-Haefeli M (1986) Was bringt die Micropille? Therapeutische Umschau 43:365–371
Mall-Haefeli M, Spornitz UM (1984) Das Progestasertsystem. In: Beller FK, Schweppe KW, Wagner H (eds) Intrauterinpessare. Edition Medizin Weinheim, pp 279–293
Mall-Haefeli M, Ludwig KS, Spornitz UM, Uettwiller A (1976) Die Low-Dosis-Gestagen-Therapie. Geburtsh u Frauenheilk 36:645–660
Mall-Haefeli M, Werner-Zodrow I, Uettwiller A (1977) Vergleichende Hormonprofile an Frauen vor und nach der Einlage eines intrauterinen Progesteron Systems. First European Symposium on the Intrauterine Progesterone-releasing contraceptive Systems 18 March 1977
Marchand F (1904) Beitrag zur Kenntnis der normalen und pathologischen Histologie der Dezidua. Arch Gynaek 72:155–167
Martinez-Manautou J (1975) Contraception by intrauterine release of progesterone. J Steroid Biochem 6:889–894
Maslar IA, Riddick DH (1979) Prolactin production by human endometrium during the normal menstrual cycle. Am J Obstet Gynecol 135:751–754
Mayor HD, Hampton JC, Rosario B (1961) A simple method for removing the resin from epoxy-embedded tissue. J Cell Biol 9:909–910
McRae MA, Newman GR, Walker SM, Jasani B (1986) Immunohistochemical identification of prolactin and 24K protein in secretory endometrium. Fertil Steril 45:643–648
Merker HJ, Herbst R, Kloss K (1968) Elektronenmikroskopische Untersuchungen an den Mitochondrien des menschlichen Uterusepithels waehrend der Sekretionsphase. Z Zellforsch 86:139–152
Meuris S, Soumenkoff G, Malengreau A, Robyn C (1980) Immunoenzymatic localization of prolactin-like immunoreactivity in decidual cells of the endometrium from pregnant and nonpregnant women. J Histochem Cytochem 28:1347–1350
Meyer R (1911) Ueber Corpus Luteum-Bildung beim Menschen. Arch Gynaekol 93:354–404
Michna H (1988) Intracellular collagen fibrils: evidence of an intracellular source from experiments with tendon fibroblasts and fibroblastic tumour cells. J Anatomy 158:1–12
Miller MM, O'Morchoe CC (1982) Decidual cell reaction induced by prostaglandin F 2 alpha in the mature oophorectomized rat. Cell Tissue Res 225:189–199
Monneron A, Bernhard W (1966) Action de certaines enzymes sur des tissues inclus en epon. J Microsop 5:697–713
Moon-Il C, Garant PR (1981) An electron microscopic radio-autographic study of collagen secretion in periodontal ligament fibroblasts of the mouse. II. Colchicine-treated fibroblasts Anat Rec 201:587–598
Morano E, Sirtori C, Vecchietti G (1959) Ultrastruttura dell'endometrio umano. Tumori 45:1–12
More IAR, Masterton RG (1975) The fine structure of the human endometrial ciliated cell. J Reprod Fertil 45:343–348

More IAR, McSeveney D (1980) The three-dimensional structure of the nucleolar channel system in the endometrial glandular cell. Serial sectioning and high voltage microscopic studies. J Anat 130:673–682
More IAR, Armstrong EM, McSeveney D, Chatfield WR (1974) The morphogenesis and fate of the nucleolar channel system in the human endometrial glandular cell. J Ultrastr Res 47:74–85
More IAR, Armstrong EM, McSeveney D (1975) Observations on the three-dimensional structure of the nucleolar channel system of the human endometrial glandular cell. J Anat 119:163–167
Moricard R, Moricard F (1964) Modificatons cytoplasmiques et nucleares ultrastructurales uterines au cours de l'etat follico-luteinique a glycogene massif. Gyn Obst (Paris) 63:203–219
Moyer DL, Mishell DR (1971) Reactions of human endometrium to the intrauterine foreign body. II. Long term effects on endometrial histology and cytology. Am J Obstet Gynecol 111:66–80
Moyer DL, Shaw ST (1984) Biochemische Aspekte von Intrauterinpessaren. In: Beller FK, Schweppe KW, Wagner H (eds) Intrauterinpessare. Edition Medizin Weinheim, pp 23–35
Nakao K, Meyer CJ, Noda Y (1971) Progesterone specific protein crystals in the endometrium: an electron microscopic study. Am J Obstet Gynecol 111:1034–1038
Nilsson CG, Luukkainen T, Arko H (1978) Endometrial morphology of women using a D-norgestrel-releasing intrauterine device. Fertil Steril 4:397–401
Nilsson O (1962a) Electron microscopy of the glandular epithelium in the human uterus. I. Follicular phase. J Ultrastr Res 6:413–421
Nilsson O (1962b) Electron microscopy of the glandular epithelium in the human uterus. II. Early and late luteal phase. J Ultrastr Res 6:422–431
Novat D, Laufer N, Kopolovic J (1986) Artificially induced endometrial cycles and establishment of pregnancies in the absence of ovaries. N Engl J Med 806–816
Noyes RW, Hertig AT, Rock J (1950) Dating the endometrial biopsy. Fertil Steril 1:3–25
O'Shea, JD, Kleinfeld RG, Morrow HA (1983) Ultrastructure of decidualization in the pseudopregnant rat. Am J Anat 166:271–298
Peel S, Bulmer D (1977) The fine structure of the rat metrial gland in relation to the origin of the granulated cells. J Anat 123:687–696
Peel S, Stewart I, Bulmer D (1982) The morphology of granulated metrial gland cells in lethally irradiated, bone marrow reconstituted mice. J Anat 135:849
Pharriss BB, Erickson R, Bashaw J, Hoff S, Place VA, Zaffaroni A (1974) Progestasert: an uterine therapeutic system for long term contraception: I. Philosophy and clinical efficacy. Fertil Steril 25:915–921
Pijnenborg R, Dixon G, Robertson WB, Brosens I (1980) Trophoblastic invasion of human decidua from 8 to 18 weeks of pregnancy. Placenta 1:3–19
Piotrow PT, Rinehart W, Schmidt JC (1979) IUDs – update on safety effectiveness and research. Popul Rep [B] 3:1–25
Ramsey E (1973) The history of the uterus. In: Norris HJ, Hertig AT, Abell MB (eds) The uterus. Williams and Wilkins, Baltimore, pp 1–16
Rewell RE (1972) Extra-uterine decidua. J Pathol 105:219–222
Riddick DH, Daly DC (1982) Decidual prolactin production in human gestation. Semin Perinatol 6:229–237
Riddick DH, Kusmik WF (1977) Decidua: a possible source of amniotic fluid prolactin. Am J Obstet Gynecol 127:187–190
Rock J (1963) The time has come. Longmans, Green, London
Rock J, Bartlett M (1937) Biopsy studies of human endometrium. JAMA 108:2022–2028
Romeis B (1968) Mikroskopische Technik. Oldenbourg, Muenchen
Rybo G, Bergqvist A (1976) Comparison of menstrual blood loss with the proogestasert system and the Cu-T-200. Preliminary results. Symposium on clinical experience with the progesterone uterine therapeutic system. Acapulco, Mexico
Ryman BE, Whelan WJ (1971) New aspects of glycogen metabolism. In: Nord FF (ed) Advances in Enzymology, vol 34. Wiley, New York, pp 285–443

Sananes N, Baulieu EE, Le Goascogne C (1981) A role for prostaglandins in decidualization of the rat uterus. J Endocr 89:25–33
Schmidt CL (1985) Endometriosis: a reappraisal of pathogenesis and treatment. Fertil Steril 44:157–173
Schmidt W, Pfaller K, Klima G (1982) Licht- und elektronenmikroskpische Untersuchungen an Eihäuten des Menschen. Zentralbl Gynaekol 104:1137–1148
Schmorl Dr (1987) Ueber grosszellige (decidua-aehnliche) Wucherungen auf dem Peritoneum und den Ovarien bei intrauteriner Schwangerschaft. Monatsschr Geburtsh Gynaek 5:46–50
Schroeder R (1915) Anatomische Studien zur normalen und pathologischen Physiologie des Menstruationszyklus. Arch Gynaek. 104:27–102
Schweppe KW, Kindermann G (1988) Endometriose. Klinik und Therapie. In: Friedberg V, Thomsen K (eds) Gynaekologie und Geburtshilfe. (Spezielle Gynaekologie 2 vol III/2.) Thieme, Stuttgart, pp 13.12–13.27
Schweppe KW, Wynn RM (1984) Endocrine dependency of endometriosis: an ultrastructural study. Europ J Obstet Gynec Reprod Biol 17:193–208
Schweppe KW, Wynn RM, Beller FK (1984) Ultrastructural comparison of endometriotic implants and eutopic endometrium. Am J Obstet Gynecol 148:1024–1039
Seiler BW (1832) Die Gebaermutter und das Ei des Menschen in den ersten Schwangerschaftsmonaten. Walther'sche Hofbuchhandlung, Dresden
Sengel A, Stoebner P (1970a) Ultrastructure de l'endometre humain normal II. Les glandes. Z Zellforsch 109:260–278
Sengel A, Stoebner P (1970b) Ultrastructure de l'endometre humain normal. I. Le chorion cytogene. Z Zellforsch 109:245–259
Shelesnyak M (1957) Aspects of reproduction. Some experimental studies on the mechanism of ovaimplantation in the rat. Rec Prog Horm Res 13:269–322
Shelesnyak MC, Tic L (1963) Studies on the mechanism of decidualization: IV. Acta Endocrinol 42:465–472
Shepard MK, Senturia YD (1977) Comparison of serum progesterone and endometrial biopsy for conformation of ovulation and evaluation of luteal function. Fertil Steril 28:541–548
Shimizu K, Furaya T, Takeo Y, Shirama K, Maekawa K (1983) Clearance of materials from breakdown of uterine collagen in mice during postpartum involution. Acta Anat 116:10–13
Spornitz UM (1978) Studies on the liver of Xenopus laevis. III. The ultrastructure and the glycogen content of the developing liver. Anat Embryol 154:1–25
Spornitz UM (1985) Pathophysiological and undesirable interactions between the genital tract and copper IUDs. Adv Contrac Deliv Syst 2:351–353
Spornitz UM (1987) 3-D-reconstruction of the nucleolar channel system in the human endometrial glandular cell. Acta Anat 128:345
Spornitz UM (1991) Decidualized tissue in ectopic pregnancy. Acta anat. 140:199–200
Spornitz UM. Hafez ESE (1982) Morphology of the basement membrane in carcinoma of the cervix. In: Hafez ESE, Smith JP (eds) Carcinoma of the cervix. Martinus Nijhoff, Den Haag (Developments in obstetrics and gynecology, vol 6)
Spornitz UM, Ludwig KS, Mall-Haefeli M, Werner-Zodrow I, Uettwiller A (1980) The effect of a progesterone releasing IUD on the morphology of the endometrium and the oviduct. In: Hafez ESE, Van Os WAA (eds) Medicated intrauterine devices. pp 44–54 (Developments in obstetrics and Gynecology, vol 5)
Spornitz UM, Ludwig KS, Mall-Haefeli M (1983) Ultrastructure of the decidual response to a progesterone-releasing IUD. In: Hasson H, Hafez ESE, Van Os WAA (eds) Biomedical Aspects of IUDs. MTP Press Lancaster, pp 3–23 (Advances in reproductive health care, vol 5)
Spornitz UM, Makabe S, Hafez ESE, Van Os WAA (1984) Scanning electron microscopy of the IUD tail. Contracept Deliv Syst 5:29–45
Staubesand J (1977). Intracellular collagen in smooth muscle: the fine structure of the artificially occluded rat artery and ureter, and of human varicose and arteriosclerotic vessels. Beitr Pathol 161:187–193

Tatum HJ (1972) Intrauterine contraception. Am J Obstet Gynecol 112:1000–1023
Tauber PF, Nohlen M (1984) Blutungsstoerungen bei IUP-Traegerinnen und Behandlungsmoeglichkeiten. In: Beller FK, Schweppe KW, Wagner H (eds) Intrauterinpessare. Edition Medizin Weinheim, pp 139–152
Terzakis JA (1965) The nucleolar channel system. J Cell Biol 27:293–304
Thiery JP (1967) Mise en evidence des polysaccharides sur coupes fines en microscopie electronique. J Microscopie 6:987–1018
Toder V, Madanes A, Gleicher R (1988) Immunologic aspects of IUD action. Contraception 37:391–403
Toldt C (1877) Lehrbuch der Gewebelehre. Enke, Stuttgart
Tsibris JCM, Cazenave CR, Cantor B, Notelovitz M, Kalra PS, Spellacy WN (1978) Distribution of cytoplasmic estrogen and progesterone receptors in human endometrium. Am J Obstet Gynecol 132:449–454
Tyson JE, Hwang P, Guyda H, Friesen HG (1972) Studies of prolactin secretion in human pregnancy. Am J Obstet Gynecol 113:14–20
Ulesco-Stroganoff K (1908) Zur Frage von dem feinsten Bau des Deciduagewebes, seiner Histogenese. Bedeutung und dem Orte seiner Entwicklung im Genitalapparat der Frau. Arch Gynaek 86:542–563
Van Os WAA, De Nooyer CCA, Rhemrev PER (1980) Copper IUDs: general considerations. In: Hafez ESE, Van Os WAA (eds) Medicated intrauterine devices. Martinus Nigholf, Den Haag, Den Haag, pp 111–118 (Developments in obstetrics and Gynecology, vol 5)
Vecchietti G, Morano E (1959). Investigation of human endometrium by the electron microscope. Int J Fertil 4:109–114
Verma V (1983) Ultrastructural changes in human endometrium at different phases of the menstrual cycle and their functional significance. Gynecol Obstet Invest 15:193–212
Vesalius A (1543) De Humani corporis fabrica. Libri septem. J Oporinus, Basel
Vierikko P, Kauppila A, Roennberg L, Vihko R (1985) Steroidal regulation of endometriosis tissue: lack of induction of 17beta-hydroxysteroid dehydrogenase activity by progesterone, medroxyprogesterone acetate, or danazol. Fertil Steril 43:218–224
Von Ebner V (1902) Koelliker's Handbuch der Gewebelehre des Menschen, 6th edn: vol 3. W Engelmann, Leipzig
Von Luedinghausen M, Anastasiadis P (1984) Anatomic basis of endometrial cytology. Acta Cytol 28:555–562
Vrensen, GFJM, Kuyper CMA (1969) Involvement of rough endoplasmic reticulum and ribosomes in early stages of glycogen repletion in rat liver. J Microscopie 8:599–614
Vu Hai MT, Logeat F, Warembourg M, Milgrom E (1977) Hormonal control of progesterone receptors. Ann NY Acad Sci 286:199–209
Wadsworth PF, Lewis DJ, Heywood R (1980) The ultrastructural features of progestagen-induced decidual cells in the rhesus monkey (Macaca mulatta). Contraception 22:189–198
Wagner H (1976) Untersuchungen zum kontrazeptiven Effekt von Kupfer. Habilitationsschrift, Wilhelms-Universitaet Muenster, pp 1–226
Wagner H, Beller FK (1979) Ultrastructural aspects of human endometrium during menstruation. The biology of the fluids of the female genital tract. pp 187–201
Weibel ER (1969) Stereological methods for morphometric cytology. Int Rev Cytol 26:235–302
Weibel ER, Kistler GS, Scherle WF (1966) Practical stereological methods for morphometric cytology. J Cell Biol 30:23–38
Weill P (1921) Les cellules granuleuses des muqueuses intestinale et uterine. Arch Mikr Anat 17:77–82
Weller CV (1935) The ectopic decidual reaction and its significance in endometriosis. Am J Pathol 11:287–291
Wessel W (1959) Die menschlichen Deciduazellen und ihre "Kollageneinschluesse" im Elektronenmikroskop. Virchows Arch [A] 332:224–235
Wetzstein R (1958) Elektronenmikroskopische Studien am Endometrium. Verh Anat Ges 55:218–226

Wewer UM, Faber M, Liotta LA, Albrechtsen R (1985) Immunochemical and ultrastructural assessment of the nature of the pericellular basement membrane of human decidual cells. Lab Invest 53:624–633
Wienke EC, Cavazos F, Hall DG, Lucas FV (1968) Ultrastructure of the human endometrial stroma cell during the menstrual cycle. Am J Obstet Gynecol 102:65–77
Wilborn WH, Flowers CE (1979) Histoenzymology of human endometrium during the proliferative and secretory phases. Beller FK, Schumacher GFB (eds) The biology of the fluids of the female genital tract. Elsevier, New York, pp 73–87
Williams HE, Provine HT (1966) Effects of estradiol on glycogen synthetase in the rat uterus. Endocrinology 78:786–790
Wist A (1954) Decidual reaction and tubal pregnancy. Acta Obstet Gynecol Scand 33:83–90
Wittlinger H (1980) Clinical effects of estrogens. In: Dallenbach-Hellweg G (ed) Functional morphologic changes in female sex organs induced by exogenous hormones. Springer, Berlin Heidelberg New York, pp 67–71
Worth ER, Lucas FV (1978) A three dimensional study of endometrial mitochondria during the menstrual cycle. Am J Obstet Gynecol 130:152–155
Wynn RM (1968) Fine structural effects of intrauterine contraceptives on the human endometrium. Fertil Steril 19:867–882
Wynn RM (1974) Ultrastructural development of the human decidua. Am J Obstet Gynecol 118:652–670
Wynn RM (1978) Ultrastructure of the endometrium and its modification by contraceptive agents. In: Ludwig H, Tauber F (eds) Human fertilization. Thieme, Stuttgart, pp 215–236
Wynn RM, Harris JA (1967) Ultrastructural cyclic changes in the human endometrium. I. Normal preovulatory phase. Fertil Steril 18:632–648
Wynn RM, Woolley RS (1967) Ultrastructural cyclic changes in the human endometrium. Fertil Steril 18:721–738
Yasuzumi G, Yabumoto N, Saito K, Tsubo I (1981) In vivo production of nucleolar channel system in human endocervical secretory cells. J Submicrosc Cytol 13:639–647
Ylä-Jääski J, Kübler, O (1987a) A system for fast display of medical 3-D images. 9th Annual Conference of the Society for Engineering in Medicine and Biology, Boston
Ylä-Jääski J, Kübler O (1987b) Automatic segmentation and fast display of volume data. Proceedings 5th Scandinavian Conference Image Analysis, Stockholm, pp 375–382
Zanefeld LJD (1978) Capacitation of spermatozoa. In: Ludwig H, Tauber PF (eds) Human fertilization. Twieme, Stuttgart, pp 128–136
Zipper J (1984) Historische Entwicklung der kupferhaltigen Intrauterinpessare. In: Beller FK, Schweppe KW, Wagner H (eds) Intrauterinpessare. Edition Medizin, Weinheim, pp 1–10
Zipper JA, Tatum HJ, Pastene L, Rivera M (1969) Contraception through the use of intrauterine metals. I. Copper as an adjunct to the "T" device. Am J Obstet Gynecol 105:1274–1278

Subject Index

Advances in Anatomy, Embryology and Cell Biology

Editors: F. Beck, W. Hild, W. Kriz, J. E. Pauly, Y. Sano, T. H. Schiebler

Volume 123

K. E. Åström, H. deF. Webster,
Bethesda, MD

The Early Development of the Neopallial Wall and Area Choroidea in Fetal Rats

A Light and Electron Microscopic Study

1991. VI, 83 pp. 32 figs.
Softcover ISBN 3-540-53910-7

This is the first monograph to deal with the early pre-neural period of brain development. It includes original research and a review of the literature and will interest all scientists who investigate the development of nerve cells and gene regulation. This work will improve our understanding of many developmental abnormalities of the nervous system.

Volume 122

K. Yu. Reznikov, University of Moscow

Cell Proliferation and Cytogenesis in the Mouse Hippocampus

1991. VI, 83 pp. 30 figs. 7 tabs.
Softcover ISBN 3-540-53689-2

This research monograph reviews the results of the study of cell proliferation, cell death, neurogenesis and gliogenesis in the mouse hippocampus. The book presents original maps of distribution of mitoses and pyknoses in the developing Ammon's horn and dentate gyrus. It also gives an analysis of the location, age dynamics and origin of proliferating and dying cells. Data is given on how cell composition is formed. In the concluding section, the specific features of neurogenesis in the hippocampus and their possible relation to learning and memory processes are discussed.

Volume 121

P. H. M. F. van Domburg, H. J. ten Donkelaar,
University of Nijmegen

The Human Substantia Nigra and Ventral Tegmental Area

A Neuroanatomical Study with Notes on Aging and Aging Diseases

1991. X, 132 pp. 38 figs. 4 tabs.
Softcover ISBN 3-540-52823-7

This book provides a comprehensive survey of the structure and fiber connections of the human midbrain, specifically of the substantia nigra and ventral tegmental area. The cellular and chemical architecture of these structures is analyzed and their fiber connections are discussed.
The role they play in degenerative diseases of the nervous system, such as Alzheimer's and Parkinson's diseases, is evaluated. Some functional and pathophysiological considerations are included.

Volume 120

L. J. Wurzinger, Technical University of Munich

Histophysiology of the Circulating Platelet

1990. VII, 96 pp. 42 figs. 9 tabs.
Softcover ISBN 3-540-52258-1

This volume closes the gap between knowledge of the platelet based on in vitro studies and knowledge of platelet hemostatic and thrombogenic function in vivo. An exhaustive review of the relevant literature, including recent ultrastructural and cell-biological studies, forms the bridge between basic research concepts and their implications for platelet function in the flowing blood.

Springer-Verlag
Berlin
Heidelberg
New York
London
Paris
Tokyo
Hong Kong
Barcelona
Budapest

GPSR Compliance

The European Union's (EU) General Product Safety Regulation (GPSR) is a set of rules that requires consumer products to be safe and our obligations to ensure this.

If you have any concerns about our products, you can contact us on ProductSafety@springernature.com

In case Publisher is established outside the EU, the EU authorized representative is:

Springer Nature Customer Service Center GmbH
Europaplatz 3
69115 Heidelberg, Germany

Batch number: 10704763

Printed by Printforce, the Netherlands